Effectiveness and Impact of Corporate Average Fuel Economy (CAFE) Standards

Committee on the Effectiveness and Impact of Corporate
Average Fuel Economy (CAFE) Standards

Board on Energy and Environmental Systems
Division on Engineering and Physical Sciences

Transportation Research Board

National Research Council

NATIONAL ACADEMY PRESS
Washington, D.C.

NATIONAL ACADEMY PRESS   2101 Constitution Avenue, N.W.   Washington, DC 20418

NOTICE: The project that is the subject of this report was approved by the Governing Board of the National Research Council, whose members are drawn from the councils of the National Academy of Sciences, the National Academy of Engineering, and the Institute of Medicine. The members of the committee responsible for the report were chosen for their special competences and with regard for appropriate balance.

This report and the study on which it is based were supported by Grant No. DTNH22-00-G-02307. Any opinions, findings, conclusions, or recommendations expressed in this publication are those of the author(s) and do not necessarily reflect the views of the organizations or agencies that provided support for the project.

Library of Congress Control Number: 2001097714
International Standard Book Number: 0-309-07601-3

Available in limited supply from:

Board on Energy and Environmental Systems
National Research Council
2101 Constitution Avenue, N.W.
HA-270
Washington, DC 20418
202-334-3344

Additional copies are available for sale from:

National Academy Press
2101 Constitution Avenue, N.W.
Box 285
Washington, DC 20055
800-624-6242 or 202-334-3313
(in the Washington metropolitan area)
http://www.nap.edu

Copyright 2002 by the National Academy of Sciences. All rights reserved.

Printed in the United States of America

THE NATIONAL ACADEMIES

National Academy of Sciences
National Academy of Engineering
Institute of Medicine
National Research Council

The **National Academy of Sciences** is a private, nonprofit, self-perpetuating society of distinguished scholars engaged in scientific and engineering research, dedicated to the furtherance of science and technology and to their use for the general welfare. Upon the authority of the charter granted to it by the Congress in 1863, the Academy has a mandate that requires it to advise the federal government on scientific and technical matters. Dr. Bruce M. Alberts is president of the National Academy of Sciences.

The **National Academy of Engineering** was established in 1964, under the charter of the National Academy of Sciences, as a parallel organization of outstanding engineers. It is autonomous in its administration and in the selection of its members, sharing with the National Academy of Sciences the responsibility for advising the federal government. The National Academy of Engineering also sponsors engineering programs aimed at meeting national needs, encourages education and research, and recognizes the superior achievements of engineers. Dr. Wm. A. Wulf is president of the National Academy of Engineering.

The **Institute of Medicine** was established in 1970 by the National Academy of Sciences to secure the services of eminent members of appropriate professions in the examination of policy matters pertaining to the health of the public. The Institute acts under the responsibility given to the National Academy of Sciences by its congressional charter to be an adviser to the federal government and, upon its own initiative, to identify issues of medical care, research, and education. Dr. Kenneth I. Shine is president of the Institute of Medicine.

The **National Research Council** was organized by the National Academy of Sciences in 1916 to associate the broad community of science and technology with the Academy's purposes of furthering knowledge and advising the federal government. Functioning in accordance with general policies determined by the Academy, the Council has become the principal operating agency of both the National Academy of Sciences and the National Academy of Engineering in providing services to the government, the public, and the scientific and engineering communities. The Council is administered jointly by both Academies and the Institute of Medicine. Dr. Bruce M. Alberts and Dr. Wm. A. Wulf are chairman and vice chairman, respectively, of the National Research Council.

COMMITTEE ON THE EFFECTIVENESS AND IMPACT OF CORPORATE AVERAGE FUEL ECONOMY (CAFE) STANDARDS

PAUL R. PORTNEY, *Chair*, Resources for the Future, Washington, D.C.
DAVID L. MORRISON, *Vice Chair*, U.S. Nuclear Regulatory Commission (retired), Cary, North Carolina
MICHAEL M. FINKELSTEIN, Michael Finkelstein & Associates, Washington, D.C.
DAVID L. GREENE, Oak Ridge National Laboratory, Knoxville, Tennessee
JOHN H. JOHNSON, Michigan Technological University, Houghton, Michigan
MARYANN N. KELLER, priceline.com (retired), Greenwich, Connecticut
CHARLES A. LAVE, University of California (emeritus), Irvine
ADRIAN K. LUND, Insurance Institute for Highway Safety, Arlington, Virginia
PHILLIP S. MYERS, NAE,[1] University of Wisconsin, Madison (emeritus)
GARY W. ROGERS, FEV Engine Technology, Inc., Auburn Hills, Michigan
PHILIP R. SHARP, Harvard University, Cambridge, Massachusetts
JAMES L. SWEENEY, Stanford University, Stanford, California
JOHN J. WISE, NAE, Mobil Research and Development Corporation (retired), Princeton, New Jersey

Project Staff

JAMES ZUCCHETTO, Director, Board on Energy and Environmental Systems (BEES)
ALAN CRANE, Responsible Staff Officer, Effectiveness and Impact of Corporate Average Fuel Economy (CAFE) Standards
STEPHEN GODWIN, Director, Studies and Information Services (SIS), Transportation Research Board (TRB)
NANCY HUMPHREY, Senior Program Officer, SIS, TRB
PANOLA D. GOLSON, Senior Project Assistant, BEES
ANA-MARIA IGNAT, Project Assistant, BEES

Editor

DUNCAN BROWN

---

[1] NAE = member, National Academy of Engineering

BOARD ON ENERGY AND ENVIRONMENTAL SYSTEMS

ROBERT L. HIRSCH, *Chair*, Advanced Power Technologies, Inc., Washington, D.C.
RICHARD E. BALZHISER, NAE,[1] Electric Power Research Institute, Inc. (retired), Menlo Park, California
DAVID L. BODDE, University of Missouri, Kansas City
PHILIP R. CLARK, NAE, GPU Nuclear Corporation (retired), Boonton, New Jersey
WILLIAM L. FISHER, NAE, University of Texas, Austin
CHRISTOPHER FLAVIN, Worldwatch Institute, Washington, D.C.
HAROLD FORSEN, NAE, Foreign Secretary, National Academy of Engineering, Washington, D.C.
WILLIAM FULKERSON, Oak Ridge National Laboratory (retired) and University of Tennessee, Knoxville
MARTHA A. KREBS, California Nano Systems Institute, Alexandria, Virginia
GERALD L. KULCINSKI, NAE, University of Wisconsin, Madison
EDWARD S. RUBIN, Carnegie Mellon University, Pittsburgh, Pennsylvania
ROBERT W. SHAW JR., Aretê Corporation, Center Harbor, New Hampshire
JACK SIEGEL, Energy Resources International, Inc., Washington, D.C.
ROBERT SOCOLOW, Princeton University, Princeton, New Jersey
KATHLEEN C. TAYLOR, NAE, General Motors Corporation, Warren, Michigan
JACK WHITE, The Winslow Group, LLC, Fairfax, Virginia
JOHN J. WISE, NAE, Mobil Research and Development Corporation (retired), Princeton, New Jersey

Staff

JAMES ZUCCHETTO, Director
RICHARD CAMPBELL, Program Officer
ALAN CRANE, Program Officer
MARTIN OFFUTT, Program Officer
SUSANNA CLARENDON, Financial Associate
PANOLA D. GOLSON, Project Assistant
ANA-MARIA IGNAT, Project Assistant
SHANNA LIBERMAN, Project Assistant

---

[1] NAE = member, National Academy of Engineering

Acknowledgments

The Committee on the Effectiveness and Impact of Corporate Average Fuel Economy (CAFE) Standards was aided by the following consultants: Tom Austin, Sierra Research, Inc.; K.G. Duleep, Energy and Environmental Analysis, Inc.; and Steve Plotkin, Argonne National Laboratory. These consultants provided analyses to the committee, which the committee used in addition to the many other sources of information it received.

This report has been reviewed by individuals chosen for their diverse perspectives and technical expertise, in accordance with procedures approved by the Report Review Committee of the National Research Council (NRC). The purpose of this independent review is to provide candid and critical comments that will assist the authors and the NRC in making the published report as sound as possible and to ensure that the report meets institutional standards for objectivity, evidence, and responsiveness to the study charge. The content of the review comments and draft manuscript remain confidential to protect the integrity of the deliberative process. We wish to thank the following individuals for their participation in the review of this report:

William Agnew (NAE), General Motors Research Laboratories (retired);
Lewis Branscomb (NAS, NAE), Harvard University (emeritus);
David Cole, Environmental Research Institute of Michigan;
Kennerly H. Digges, George Washington University;
Theodore H. Geballe (NAS), Stanford University (emeritus);
Paul J. Joskow, Massachusetts Institute of Technology;
James A. Levinsohn, University of Michigan, Ann Arbor;
James J. MacKenzie, World Resources Institute; and
Marc Ross, University of Michigan, Ann Arbor.

Although the reviewers listed above have provided many constructive comments and suggestions, they were not asked to endorse the conclusions or recommendations, nor did they see the final draft of the report before its release. The review of this report was overseen by John Heywood (NAE), Massachusetts Institute of Technology, and Gerald P. Dinneen (NAE), Honeywell Inc. (retired). Appointed by the National Research Council, they were responsible for making certain that an independent examination of the report was carried out in accordance with institutional procedures and that all review comments were carefully considered. Responsibility for the final content of this report rests entirely with the authoring committee and the institution.

In addition, the committee reexamined its technical and economic analysis after the release of the prepublication copy in July 2001. The results of that reexamination were released in a letter report, which is also included in this report as Appendix F. The reviewers of that report are credited in Appendix F.

Contents

EXECUTIVE SUMMARY   1

1   INTRODUCTION   7
      Scope and Conduct of the Study, 11
      References, 12

2   THE CAFE STANDARDS: AN ASSESSMENT   13
      CAFE and Energy, 13
      Impacts on the Automobile Industry, 22
      Impact on Safety, 24
      References, 29

3   TECHNOLOGIES FOR IMPROVING THE FUEL ECONOMY   31
        OF PASSENGER CARS AND LIGHT-DUTY TRUCKS
      Fuel Economy Overview, 31
      Technologies for Better Fuel Economy, 35
      Estimating Potential Fuel Economy Gains and Costs, 40
      Hybrid Vehicles, 51
      Fuel Cells, 53
      References, 55
      Attachment 3A—A Technical Evaluation of Two Weight- and Engineering-Based
        Fuel-Efficiency Parameters for Cars and Light Trucks, 56

4   IMPACT OF A MORE FUEL-EFFICIENT FLEET   63
      Energy Demand and Greenhouse Gas Impact, 63
      Analysis of Cost-Efficient Fuel Economy, 64
      Potential Impacts on the Domestic Automobile Industry, 67
      Safety Implications of Future Increases in Fuel Economy, 69
      References, 78
      Attachment 4A—Life-Cycle Analysis of Automobile Technologies, 79

5   POTENTIAL MODIFICATIONS OF AND ALTERNATIVES TO CAFE   83
      Why Governmental Intervention?, 83
      Alternative Policies—Summary Description, 86
      More Complete Descriptions of the Alternatives, 88
      Analysis of Alternatives, 94
      References, 103
      Attachment 5A—Development of an Enhanced-CAFE Standard, 104

6 FINDINGS AND RECOMMENDATIONS 111
  Findings, 111
  Recommendations, 114

APPENDIXES

A Dissent on Safety Issues: Fuel Economy and Highway Safety,
   David L. Greene and Maryann Keller, 117
B Biographical Sketches of Committee Members, 125
C Presentations and Committee Activities, 128
D Statement of Work: Effectiveness and Impact of CAFE Standards, 130
E Acronyms and Abbreviations, 131
F Letter Report: Technology and Economic Analysis in the Prepublication Report
   Effectiveness and Impact of Corporate Average Fuel Economy (CAFE) Standards, 133

Tables and Figures

TABLES

2-1 Change in Death or Injury Rates for 100-lb Weight Reduction in Average Car or Average Light Truck, 26
2-2 Occupant Deaths per Million Registered Vehicles 1 to 3 Years Old, 28
2-3 Distribution of Motor Vehicle Crash Fatalities in 1993 and 1999 by Vehicle and Crash Type, 29

3-1 Fuel Consumption Technology Matrix—Passenger Cars, 42
3-2 Fuel Consumption Technology Matrix—SUVs and Minivans, 43
3-3 Fuel Consumption Technology Matrix—Pickup Trucks, 44
3-4 Estimated Fuel Consumption (FC), Fuel Economy (FE), and Incremental Costs of Product Development, 45
3-5 Published Data for Some Hybrid Vehicles, 52

4-1 Key Assumptions of Cost-Efficient Analysis for New Car and Light Truck Fuel Economy Estimates Using Path 3 Technologies and Costs, 65
4-2 Case 1: Cost-Efficient Fuel Economy (FE) Analysis for 14-Year Payback, 67
4-3 Case 2: Cost-Efficient Fuel Economy (FE) Analysis for 3-Year Payback, 67
4-4 Relative Collision Claim Frequencies for 1998–2000 Models, 73

4A-1 Vehicle Architecture and Fuels Used in the MIT and General Motors et al. Studies, 79

5-1 Incentives of the Various Policy Instruments for Seven Types of Fuel Use Response, 95
5-2 Issues of Cost Minimization for the Various Policy Instruments, 98
5-3 Performance Trade-offs for the Various Policy Instruments, 99

A-1 Estimated Effects of a 10 Percent Reduction in the Weights of Passenger Cars and Light Trucks, 120

FIGURES

2-1 Oil price shocks and economic growth, 1970–1999, 14
2-2 Automotive fuel economy standards (AFES) and manufacturers' CAFE levels, 14
2-3 Average weights of domestic and imported vehicles, 15
2-4 Fleet fuel economy of new and on-road passenger cars and light trucks, 16
2-5 Passenger car size and weight, 17

2-6   Trends in fuel-economy-related attributes of passenger cars, 1975–2000, 17
2-7   Trends in fuel-economy-related attributes of light trucks, 1975–2000, 17
2-8   Average new car price and fuel economy, 18
2-9   Passenger car and light-truck travel and fuel use, 19
2-10 Employment and productivity in the U.S. automotive industry, 22
2-11 Net profit rates of domestic manufacturers, 1972–1997, 22
2-12 Investments in retooling by domestic automobile manufacturers, 1972–1997, with automotive fuel economy standards (AFES) for passenger cars and trucks, 23
2-13 R&D investments by domestic automobile manufacturers, 1972–1997, with automotive fuel economy standards (AFES) for passenger cars and trucks, 24
2-14 Motor vehicle crash death rates, 1950–1998, 25

3-1   Energy use in vehicles, 32
3-2   Where the energy in the fuel goes, 33
3-3   EPA data for fuel economy for MY 2000 and 2001 cars and light trucks, 34
3-4   Subcompact cars. Incremental cost as a function of fuel consumption, 46
3-5   Compact cars. Incremental cost as a function of fuel consumption, 46
3-6   Midsize cars. Incremental cost as a function of fuel consumption, 47
3-7   Large cars. Incremental cost as a function of fuel consumption, 47
3-8   Small SUVs. Incremental cost as a function of fuel consumption, 48
3-9   Midsize SUVs. Incremental cost as a function of fuel consumption, 48
3-10 Large SUVs. Incremental cost as a function of fuel consumption, 49
3-11 Minivans. Incremental cost as a function of fuel consumption, 49
3-12 Small pickups. Incremental cost as a function of fuel consumption, 50
3-13 Large pickups. Incremental cost as a function of fuel consumption, 50
3-14 Relationship between the power of an internal combustion engine and the power of an electric motor in a hybrid electric vehicle, 51
3-15 Breakdown of fuel economy improvements by technology combination, 52
3-16 Working principles of a PEM fuel cell, 53
3-17 State of the art and future targets for fuel cell development, 54
3-18 Typical fuel cell efficiency, 55

3A-1 Dependence of fuel consumption on fuel economy, 56
3A-2 Weight-specific fuel consumption versus weight for all vehicles, 57
3A-3 Fleet fuel economy, 57
3A-4 Best-in-class fuel-efficiency analysis of 2000 and 2001 vehicles, 58
3A-5 LSFC versus payload for a variety of vehicles, 59
3A-6 Fuel consumption versus payload, 59
3A-7 Payload versus LSFC, 60
3A-8 Payload for a variety of vehicles, 60
3A-9 Fuel economy as a function of average WSFC for different classes of vehicles, 61
3A-10 Fuel economy versus average payload for different classes of vehicles, 62

4-1   Fuel use in alternative 2013 fuel economy scenarios, 63
4-2   Fuel savings of alternative 2013 fuel economy improvement targets, 64
4-3   Fuel-cycle greenhouse gas emissions in alternative 2013 fuel economy cases, 64
4-4   Greenhouse gas emissions reductions from hypothetical alternative fuel economy improvements targets, 64
4-5   Passenger car fuel economy cost curves from selected studies, 68
4-6   Light-truck fuel economy cost curves from selected studies, 68
4-7   Occupant death rates in single-vehicle crashes for 1990–1996 model passenger vehicles by weight of vehicle, 71
4-8   Occupant death rates in two-vehicle crashes for 1990–1996 model passenger vehicles by weight of vehicle, 71

TABLES AND FIGURES

4-9   Occupant death rates in other vehicles in two-vehicle crashes for 1990–1996 model passenger vehicles, 72
4-10 Pedestrian/bicyclist/motorcyclist death rates for 1990–1996 model passenger vehicles by vehicle weight, 72

4A-1 Life-cycle comparisons of technologies for midsize passenger vehicles, 80
4A-2 Well-to-wheels total system energy use for selected fuel/vehicle pathways, 81
4A-3 Well-to-wheels greenhouse gas emissions for selected fuel/vehicle pathways, 82

5-1   The operation of the current CAFE standards: passenger cars, gasoline engines only, 1999, 92
5-2   Fuel economy targets under the Enhanced-CAFE system: cars with gasoline engines, 93

5A-1 Gallons used per 100 miles (cars only, gasoline engines only), 104
5A-2 Regression line through the car data in Figure 5A-1 (passenger cars only, gasoline engines only), 105
5A-3 Gallons to drive 100 miles with regression lines (cars and trucks, gasoline engines only), 106
5A-4 Gallons used per 100 miles (all vehicles), 106
5A-5 Weight-specific fuel consumption, 107
5A-6 Enhanced CAFE targets, 109
5A-7 Enhanced CAFE targets in WSFC units, 109

A-1   NHTSA passenger-side crash ratings for MY 2001 passenger cars, 121
A-2   NHTSA driver-side crash ratings for MY 2001 passenger cars, 121
A-3   Estimated frequency of damage to a tree or pole given a single-vehicle crash with a fixed object, 122
A-4   NHTSA static stability factor vs. total weight for MY 2001 vehicles, 123
A-5   Traffic fatality rates and on-road light-duty miles per gallon 1996–2000, 123

Executive Summary

In the wake of the 1973 oil crisis, the U.S. Congress passed the Energy Policy and Conservation Act of 1975, with the goal of reducing the country's dependence on foreign oil. Among other things, the act established the Corporate Average Fuel Economy (CAFE) program, which required automobile manufacturers to increase the sales-weighted average fuel economy of the passenger car and light-duty truck fleets sold in the United States. Today, the light-duty truck fleet includes minivans, pickups, and sport utility vehicles. Congress itself set the standards for passenger cars, which rose from 18 miles per gallon (mpg) in automobile model year (MY) 1978 to 27.5 mpg in MY 1985. As authorized by the act, the Department of Transportation (DOT) set standards for light trucks for model years 1979 through 2002. The standards are currently 27.5 mpg for passenger cars and 20.7 mpg for light trucks. Provisions in DOT's annual appropriations bills since fiscal year 1996 have prohibited the agency from changing or even studying CAFE standards.

In legislation for fiscal year 2001, Congress requested that the National Academy of Sciences, in consultation with the Department of Transportation, conduct a study to evaluate the effectiveness and impacts of CAFE standards.[1] In particular, it asked that the study examine the following, among other factors:

1. The statutory criteria (economic practicability, technological feasibility, need for the United States to conserve energy, the classification definitions used to distinguish passenger cars from light trucks, and the effect of other regulations);
2. The impact of CAFE standards on motor vehicle safety;
3. Disparate impacts on the U.S. automotive sector;
4. The effect on U.S. employment in the automotive sector;
5. The effect on the automotive consumer; and
6. The effect of requiring separate CAFE calculations for domestic and nondomestic fleets.

In response to this request, the National Research Council (NRC) established the Committee on the Effectiveness and Impact of Corporate Average Fuel Economy (CAFE) Standards. In consultation with DOT, the NRC developed a statement of work for the committee. The committee's work was to emphasize recent experience with CAFE standards, the impact of possible changes, and the stringency and/or structure of the CAFE program in future years. The committee held its first meeting in early February 2001. In effect, since the congressional appropriations language asked for the report by July 1, 2001, the committee had less than 5 months (from February to late June) to complete its analysis and prepare a report for the National Research Council's external report review process. In its findings and recommendations, the committee has noted where analysis is limited and further study is needed.

Following the release of the prepublication copy of this report in July 2001, the committee reviewed its technical and economic analyses. Several changes were made to the results, as reported in a letter report released in January 2002, which is reprinted in Appendix F below. These changes have been incorporated in this report also.

The CAFE program has been controversial since its inception. Sharp disagreements exist regarding the effects of the program on the fuel economy of the U.S. vehicle fleet, the current mix of vehicles in that fleet, the overall safety of passenger vehicles, the health of the domestic automobile industry, employment in that industry, and the well-being of consumers. It is this set of concerns that the committee was asked to address.

These concerns are also very much dependent on one another. For example, if fuel economy standards were raised,

---

[1]Conference Report on H.R. 4475, Department of Transportation and Related Agencies Appropriations Act, 2001. Report 106-940, as published in the *Congressional Record*, October 5, 2000, pp. H8892-H9004.

the manner in which automotive manufacturers respond would affect the purchase price, attributes, and performance of their vehicles. For this reason, the mix of vehicles that a given manufacturer sells could change, perhaps resulting in a greater proportion of smaller and lighter vehicles; this, in turn, could have safety implications, depending on the eventual mix of vehicles that ended up on the road. If consumers are not satisfied with the more fuel-efficient vehicles, that in turn could affect vehicle sales, profits, and employment in the industry. Future effects would also depend greatly on the real price of gasoline; if it is low, consumers would have little interest in fuel-efficient vehicles. High fuel prices would have just the opposite effect. In addition, depending on the level at which fuel economy targets are set and the time the companies have to implement changes, differential impacts across manufacturers would probably occur depending on the types of vehicles they sell and their competitive position in the marketplace. Thus, understanding the impact of potential changes to CAFE standards is, indeed, a difficult and complex task.

In addition to the requirement that companies meet separate fleet averages for the automobiles and light-duty trucks they sell, there are other provisions of the CAFE program that affect manufacturers' decisions. For example, a manufacturer must meet the automobile CAFE standard separately for both its import and its domestic fleet (the two-fleet rule), where a domestic vehicle is defined as one for which at least 75 percent of its parts are manufactured in the United States. Also, CAFE credits can be earned by manufacturers that produce flexible-fuel vehicles, which can run interchangeably on gasoline or an alternative fuel, such as ethanol.

Why care about fuel economy at all? It is tempting to say that improvements in vehicle fuel economy will save money for the vehicle owner in reduced expenditures for gasoline. The extent of the annual saving will depend on the level of improvement in the fuel economy (in miles per gallon of gasoline), the price of gasoline, and the miles traveled per year, as well as on the higher cost of the vehicle attributable to the fuel economy improvement. While a strong argument can be made that such savings or costs are economically relevant, that is not by itself a strong basis for public policy intervention. Consumers have a wide variety of opportunities to exercise their preference for a fuel-efficient vehicle if that is an important attribute to them. Thus, according to this logic, there is no good reason for the government to intervene in the market and require new light-duty vehicles to achieve higher miles per gallon or to take other policy measures designed to improve the fuel economy of the fleet.

There are, however, other reasons for the nation to consider policy interventions of some sort to increase fuel economy. The most important of these, the committee believes, is concern about the accumulation in the atmosphere of so-called greenhouse gases, principally carbon dioxide. Continued increases in carbon dioxide emissions are likely to further global warming. Concerns like those about climate change are not normally reflected in the market for new vehicles. Few consumers take into account the environmental costs that the use of their vehicle may occasion; in the parlance of economics, this is a classic negative externality.

A second concern is that petroleum imports have been steadily rising because of the nation's increasing demand for gasoline without a corresponding increase in domestic supply. The demand for gasoline has been exacerbated by the increasing sales of light trucks, which have lower fuel economy than automobiles. The high cost of oil imports poses two risks: downward pressure on the strength of the dollar (which drives up the costs of goods that Americans import) and an increase in U.S. vulnerability to macroeconomic shocks that cost the economy considerable real output. Some experts argue that these vulnerabilities are another form of externality that vehicle purchasers do not factor into their decisions but that can represent a true and significant cost to society. Other experts take a more skeptical view, arguing instead that the macroeconomic difficulties of the 1970s (high unemployment coupled with very high inflation and interest rates) were due more to unenlightened monetary policy than to the inherent difficulties associated with high oil prices. Most would agree that reducing our nation's oil import bill would have favorable effects on the terms of trade, and that this is a valid consideration in deliberations about fuel economy.

The committee believes it is critically important to be clear about the reasons for considering improved fuel economy. Moreover, and to the extent possible, it is useful to try to think about how much it is worth to society in dollar terms to reduce emissions of greenhouse gases (by 1 ton, say) and reduce dependence on imported oil (say, by 1 barrel). If it is possible to assign dollar values to these favorable effects (no mean feat, the committee acknowledges), it becomes possible to make at least crude comparisons between the beneficial effects of measures to improve fuel economy on the one hand, and the costs (both out-of-pocket and more subtle) on the other.

In conducting its study, the committee first assessed the impact of the current CAFE system on reductions in fuel consumption, on emissions of greenhouse gases, on safety, and on impacts on the industry (see Chapters 1 and 2). To assess the potential impacts of modified standards, the committee examined opportunities offered by the application of existing (production-intent) or emerging technologies, estimated the costs of such improvements, and examined the lead times that would typically be required to introduce such vehicle changes (see Chapter 3). The committee reviewed many sources of information on technologies and the costs of improvements in fuel economy; these sources included presentations at its meetings and available reports. It also used consultants under its direction to facilitate its work under the tight time constraints of the study. Some of the consultants' work provided analyses and information that helped the committee better understand the nature of previous fuel economy analyses. In the end, however, the committee conducted its own analyses, informed by the work of the consultants, the technical literature, and presentations at its meetings, as well as the expertise and judgment of its members, to arrive at its own range of estimates of fuel economy improvements and associated costs. Based on these analyses, the implications of modified CAFE standards are presented in Chapter 4, along with an analysis of what the committee calls cost-efficient fuel economy levels. The committee also examined the stringency and structure of the current CAFE system, and it assessed possible modifications to it, as well as alternative approaches to achieving higher fuel economy for passenger vehicles, which resulted in suggestions for improved policy instruments (see Chapter 5).

FINDINGS

Finding 1. The CAFE program has clearly contributed to increased fuel economy of the nation's light-duty vehicle fleet during the past 22 years. During the 1970s, high fuel prices and a desire on the part of automakers to reduce costs by reducing the weight of vehicles contributed to improved fuel economy. CAFE standards reinforced that effect. Moreover, the CAFE program has been particularly effective in keeping fuel economy above the levels to which it might have fallen when real gasoline prices began their long decline in the early 1980s. Improved fuel economy has reduced dependence on imported oil, improved the nation's terms of trade, and reduced emissions of carbon dioxide, a principal greenhouse gas, relative to what they otherwise would have been. If fuel economy had not improved, gasoline consumption (and crude oil imports) would be about 2.8 million barrels per day greater than it is, or about 14 percent of today's consumption.

Finding 2. Past improvements in the overall fuel economy of the nation's light-duty vehicle fleet have entailed very real, albeit indirect, costs. In particular, all but two members of the committee concluded that the downweighting and downsizing that occurred in the late 1970s and early 1980s, some of which was due to CAFE standards, probably resulted in an additional 1,300 to 2,600 traffic fatalities in 1993.[2] In addition, the diversion of carmakers' efforts to improve fuel economy deprived new-car buyers of some amenities they clearly value, such as faster acceleration, greater carrying or towing capacity, and reliability.

Finding 3. Certain aspects of the CAFE program have not functioned as intended:

- The distinction between a car for personal use and a truck for work use/cargo transport has broken down, initially with minivans and more recently with sport utility vehicles (SUVs) and cross-over vehicles. The car/truck distinction has been stretched well beyond the original purpose.
- The committee could find no evidence that the two-fleet rule distinguishing between domestic and foreign content has had any perceptible effect on total employment in the U.S. automotive industry.
- The provision creating extra credits for multifuel vehicles has had, if any, a negative effect on fuel economy, petroleum consumption, greenhouse gas emissions, and cost. These vehicles seldom use any fuel other than gasoline yet enable automakers to increase their production of less fuel efficient vehicles.

Finding 4. In the period since 1975, manufacturers have made considerable improvements in the basic efficiency of engines, drive trains, and vehicle aerodynamics. These improvements could have been used to improve fuel economy and/or performance. Looking at the entire light-duty fleet, both cars and trucks, between 1975 and 1984, the technology improvements were concentrated on fuel economy: It improved by 62 percent without any loss of performance as measured by 0–60 mph acceleration times. By 1985, light-duty vehicles had improved enough to meet CAFE standards. Thereafter, technology improvements were concentrated principally on performance and other vehicle attributes (including improved occupant protection). Fuel economy remained essentially unchanged while vehicles became 20 percent heavier and 0–60 mph acceleration times became, on average, 25 percent faster.

Finding 5. Technologies exist that, if applied to passenger cars and light-duty trucks, would significantly reduce fuel consumption within 15 years. Auto manufacturers are already offering or introducing many of these technologies in other markets (Europe and Japan, for example), where much higher fuel prices ($4 to $5/gal) have justified their development. However, economic, regulatory, safety, and consumer-preference-related issues will influence the extent to which these technologies are applied in the United States.

Several new technologies such as advanced lean exhaust gas aftertreatment systems for high-speed diesels and direct-injection gasoline engines, which are currently under development, are expected to offer even greater potential for reductions in fuel consumption. However, their development cycles as well as future regulatory requirements will influence if and when these technologies penetrate deeply into the U.S. market.

The committee conducted a detailed assessment of the

---

[2]A dissent by committee members David Greene and Maryann Keller on the impact of downweighting and downsizing is contained in Appendix A. They believe that the level of uncertainty is much higher than stated and that the change in the fatality rate due to efforts to improve fuel economy may have been zero. Their dissent is limited to the safety issue alone.

technological potential for improving the fuel efficiency of 10 different classes of vehicles, ranging from subcompact and compact cars to SUVs, pickups, and minivans. In addition, it estimated the range in incremental costs to the consumer that would be attributable to the application of these engine, transmission, and vehicle-related technologies.

Chapter 3 presents the results of these analyses as curves that represent the incremental benefit in fuel consumption versus the incremental cost increase over a defined baseline vehicle technology. Projections of both incremental costs and fuel consumption benefits are very uncertain, and the actual results obtained in practice may be significantly higher or lower than shown here. Three potential development paths are chosen as examples of possible product improvement approaches, which illustrate the trade-offs auto manufacturers may consider in future efforts to improve fuel efficiency.

Assessment of currently offered product technologies suggests that light-duty trucks, including SUVs, pickups, and minivans, offer the greatest potential to reduce fuel consumption on a total-gallons-saved basis.

Finding 6. In an attempt to evaluate the economic trade-offs associated with the introduction of existing and emerging technologies to improve fuel economy, the committee conducted what it called cost-efficient analysis. That is, the committee identified packages of existing and emerging technologies that could be introduced over the next 10 to 15 years that would improve fuel economy up to the point where further increases in fuel economy would not be reimbursed by fuel savings. The size, weight, and performance characteristics of the vehicles were held constant. The technologies, fuel consumption estimates, and cost projections described in Chapter 3 were used as inputs to this cost-efficient analysis.

These cost-efficient calculations depend critically on the assumptions one makes about a variety of parameters. For the purpose of calculation, the committee assumed as follows: (1) gasoline is priced at $1.50/gal, (2) a car is driven 15,600 miles in its first year, after which miles driven declines at 4.5 percent annually, (3) on-the-road fuel economy is 15 percent less than the Environmental Protection Agency's test rating, and (4) the added weight of equipment required for future safety and emission regulations will exact a 3.5 percent fuel economy penalty.

One other assumption is required to ascertain cost-efficient technology packages—the horizon over which fuel economy gains ought to be counted. Under one view, car purchasers consider fuel economy over the entire life of a new vehicle; even if they intend to sell it after 5 years, say, they care about fuel economy because it will affect the price they will receive for their used car. Alternatively, consumers may take a shorter-term perspective, not looking beyond, say, 3 years. This latter view, of course, will affect the identification of cost-efficient packages because there will be many fewer years of fuel economy savings to offset the initial purchase price.

The full results of this analysis are presented in Chapter 4. To provide one illustration, however, consider a midsize SUV. The current sales-weighted fleet fuel economy average for this class of vehicle is 21 mpg. If consumers consider only a 3-year payback period, fuel economy of 22.7 mpg would represent the cost-efficient level. If, on the other hand, consumers take the full 14-year average life of a vehicle as their horizon, the cost-efficient level increases to 28 mpg (with fuel savings discounted at 12 percent). The longer the consumer's planning horizon, in other words, the greater are the fuel economy savings against which to balance the higher initial costs of fuel-saving technologies.

The committee cannot emphasize strongly enough that the cost-efficient fuel economy levels identified in Tables 4-2 and 4-3 in Chapter 4 are *not* recommended fuel economy goals. Rather, they are reflections of technological possibilities, economic realities, and assumptions about parameter values and consumer behavior. Given the choice, consumers might well spend their money on other vehicle amenities, such as greater acceleration or towing capacity, rather than on the fuel economy cost-efficient technology packages.

Finding 7. There is a marked inconsistency between pressing automotive manufacturers for improved fuel economy from new vehicles on the one hand and insisting on low real gasoline prices on the other. Higher real prices for gasoline—for instance, through increased gasoline taxes—would create both a demand for fuel-efficient new vehicles and an incentive for owners of existing vehicles to drive them less.

Finding 8. The committee identified externalities of about $0.30/gal of gasoline associated with the combined impacts of fuel consumption on greenhouse gas emissions and on world oil market conditions. These externalities are not necessarily taken into account when consumers purchase new vehicles. Other analysts might produce lower or higher estimates of externalities.

Finding 9. There are significant uncertainties surrounding the societal costs and benefits of raising fuel economy standards for the light-duty fleet. These uncertainties include the cost of implementing existing technologies or developing new ones; the future price of gasoline; the nature of consumer preferences for vehicle type, performance, and other features; and the potential safety consequences of altered standards. The higher the target for average fuel economy, the greater the uncertainty about the cost of reaching that target.

Finding 10. Raising CAFE standards would reduce future fuel consumption below what it otherwise would be; however, other policies could accomplish the same end at lower cost, provide more flexibility to manufacturers, or address inequities arising from the present system. Possible alternatives that appear to the committee to be superior to the current CAFE structure include tradable credits for fuel economy improvements, feebates,[3] higher fuel taxes, standards based on vehicle attributes (for example, vehicle weight, size, or payload), or some combination of these.

Finding 11. Changing the current CAFE system to one featuring tradable fuel economy credits and a cap on the price of these credits appears to be particularly attractive. It would provide incentives for all manufacturers, including those that exceed the fuel economy targets, to continually increase fuel economy, while allowing manufacturers flexibility to meet consumer preferences. Such a system would also limit costs imposed on manufacturers and consumers if standards turn out to be more difficult to meet than expected. It would also reveal information about the costs of fuel economy improvements and thus promote better-informed policy decisions.

Finding 12. The CAFE program might be improved significantly by converting it to a system in which fuel economy targets depend on vehicle attributes. One such system would make the fuel economy target dependent on vehicle weight, with lower fuel consumption targets set for lighter vehicles and higher targets for heavier vehicles, up to some maximum weight, above which the target would be weight-independent. Such a system would create incentives to reduce the variance in vehicle weights between large and small vehicles, thus providing for overall vehicle safety. It has the potential to increase fuel economy with fewer negative effects on both safety and consumer choice. Above the maximum weight, vehicles would need additional advanced fuel economy technology to meet the targets. The committee believes that although such a change is promising, it requires more investigation than was possible in this study.

Finding 13. If an increase in fuel economy is effected by a system that encourages either downweighting or the production and sale of more small cars, some additional traffic fatalities would be expected. However, the actual effects would be uncertain, and any adverse safety impact could be minimized, or even reversed, if weight and size reductions were limited to heavier vehicles (particularly those over 4,000 lb). Larger vehicles would then be less damaging (aggressive) in crashes with all other vehicles and thus pose less risk to other drivers on the road.

Finding 14. Advanced technologies—including direct-injection, lean-burn gasoline engines; direct-injection compression-ignition (diesel) engines; and hybrid electric vehicles—have the potential to improve vehicle fuel economy by 20 to 40 percent or more, although at a significantly higher cost. However, lean-burn gasoline engines and diesel engines, the latter of which are already producing large fuel economy gains in Europe, face significant technical challenges to meet the Tier 2 emission standards established by the Environmental Protection Agency under the 1990 amendments to the Clean Air Act and California's low-emission-vehicle (LEV II) standards. The major problems are the Tier 2 emissions standards for nitrogen oxides and particulates and the requirement that emission control systems be certified for a 120,000-mile lifetime. If direct-injection gasoline and diesel engines are to be used extensively to improve light-duty vehicle fuel economy, significant technical developments concerning emissions control will have to occur or some adjustments to the Tier 2 emissions standards will have to be made. Hybrid electric vehicles face significant cost hurdles, and fuel-cell vehicles face significant technological, economic, and fueling infrastructure barriers.

Finding 15. Technology changes require very long lead times to be introduced into the manufacturers' product lines. Any policy that is implemented too aggressively (that is, in too short a period of time) has the potential to adversely affect manufacturers, their suppliers, their employees, and consumers. Little can be done to improve the fuel economy of the new vehicle fleet for several years because production plans already are in place. The widespread penetration of even existing technologies will probably require 4 to 8 years. For emerging technologies that require additional research and development, this time lag can be considerably longer. In addition, considerably more time is required to replace the existing vehicle fleet (on the order of 200 million vehicles) with new, more efficient vehicles. Thus, while there would be incremental gains each year as improved vehicles enter the fleet, major changes in the transportation sector's fuel consumption will require decades.

RECOMMENDATIONS

Recommendation 1. Because of concerns about greenhouse gas emissions and the level of oil imports, it is appropriate for the federal government to ensure fuel economy levels beyond those expected to result from market forces alone. Selection of fuel economy targets will require uncertain and difficult trade-offs among environmental benefits, vehicle safety, cost, oil import dependence, and consumer preferences. The committee believes that these trade-offs rightfully reside with elected officials.

Recommendation 2. The CAFE system, or any alternative regulatory system, should include broad trading of fuel

---

[3]Feebates are taxes on vehicles achieving less than the average fuel economy coupled with rebates to vehicles achieving better than average fuel economy.

economy credits. The committee believes a trading system would be less costly than the current CAFE system; provide more flexibility and options to the automotive companies; give better information on the cost of fuel economy changes to the private sector, public interest groups, and regulators; and provide incentives to all manufacturers to improve fuel economy. Importantly, trading of fuel economy credits would allow for more ambitious fuel economy goals than exist under the current CAFE system, while simultaneously reducing the economic cost of the program.

Recommendation 3. Consideration should be given to designing and evaluating an approach with fuel economy targets that are dependent on vehicle attributes, such as vehicle weight, that inherently influence fuel use. Any such system should be designed to have minimal adverse safety consequences.

Recommendation 4. Under any system of fuel economy targets, the two-fleet rule for domestic and foreign content should be eliminated.

Recommendation 5. CAFE credits for dual-fuel vehicles should be eliminated, with a long enough lead time to limit adverse financial impacts on the automotive industry.

Recommendation 6. To promote the development of longer-range, breakthrough technologies, the government should continue to fund, in cooperation with the automotive industry, precompetitive research aimed at technologies to improve vehicle fuel economy, safety, and emissions. It is only through such breakthrough technologies that dramatic increases in fuel economy will become possible.

Recommendation 7. Because of its importance to the fuel economy debate, the relationship between fuel economy and safety should be clarified. The committee urges the National Highway Traffic Safety Administration to undertake additional research on this subject, including (but not limited to) a replication, using current field data, of its 1997 analysis of the relationship between vehicle size and fatality risk.

1

Introduction

Fuel economy is attracting public and official attention in a way not seen for almost two decades. Gasoline prices have risen sharply over the past 2 years and fluctuated unpredictably. Moreover, concerns have developed over the reliability of the gasoline supply, particularly during peak driving seasons. Evidence also continues to accumulate that global climate change must be taken seriously. U.S. cars and trucks are responsible for a nonnegligible fraction of the world's annual emissions of carbon dioxide, the most important greenhouse gas.

Is it time to require cars and trucks to achieve a higher level of fuel economy? Or do such regulations do more harm than good? These questions led Congress to request a study from the National Academy of Sciences.

This report is the result of a very short, very intense study by a committee assembled to answer these questions (see Appendix B for biographies of committee members). It is intended to help policy makers in Congress and the executive branch and those outside the government determine whether and how fuel economy standards should be changed. Insofar as possible, it assesses the impact of fuel economy regulation on vehicles, energy use, greenhouse gas emissions, automotive safety, the automotive industry, and the public.

This report is the successor to another National Research Council (NRC) report on the subject and owes a great debt to the committee that prepared that report. The earlier committee began its work in May of 1991 as the Committee on Fuel Economy of Automobiles and Light Trucks, following a request from the Federal Highway Administration and the National Highway Traffic Safety Administration. It was the charge of that committee (the fuel economy committee) to study both the feasibility and the desirability of a variety of efforts to improve the fuel economy of the light-duty vehicle fleet in the United States. More than a year later, the committee issued its report, *Automotive Fuel Economy: How Far Should We Go?* (NRC, 1992).

It is difficult to summarize neatly the conclusions of that report. Briefly, though, the fuel economy committee found in 1992 as follows:

- "Practically achievable" improvements in vehicle fuel economy were possible, and these improvements would lie between, on the one hand, what would happen with no government intervention and, on the other, the results of implementing all technologically possible efficiency-enhancing measures without regard to cost, safety, or other important factors.
- Despite considerable uncertainty on this issue, *if* downweighting was used to improve fuel economy, there would probably be an adverse effect on passenger safety, all else being equal.
- While emissions standards for new cars had obvious advantages, they could make it more difficult to improve automobile fuel economy.
- The automobile manufacturing industry, which was in a sharp downturn in 1992, could be harmed by fuel economy standards "of an inappropriate form" that increased new car prices and hurt sales, or that shifted purchases to imported vehicles.
- When gasoline prices were low, consumers had limited interest in purchasing vehicles with high fuel economy, unless those same vehicles also delivered the performance characteristics—horsepower, acceleration, options—that consumers appeared to desire.
- Finally, a variety of alternatives to the then-current corporate average fuel economy standards should be considered, including changing the form of the program, increasing the price of gasoline, and adopting a system of taxes and rebates to discourage the production of "gas guzzlers" and reward "gas sippers."

Now, nearly a decade after the 1992 study began, another NRC committee has completed its work (see Appendix C for a list of the committee's meetings and site visits). While created to look at some of the same issues as the earlier group, the Committee on the Effectiveness and Impact of Corporate Average Fuel Economy (CAFE) Standards was born of a different time and directed to address a somewhat different set of concerns. For instance, the impetus for the earlier committee was a sharp, though temporary, increase in oil and gasoline prices in the wake of the Gulf War. Despite a recent increase in oil and gasoline prices related to two factors—the renewed pricing power of the Organization of Petroleum Exporting Countries (OPEC) and capacity constraints in the domestic refining industry—no serious supply interruptions motivated this report.

Similarly, and as was reflected in its findings, the earlier committee was charged with examining a wide variety of approaches that could improve the fuel economy of the passenger vehicle fleet, including changes in required fuel economy standards, increases in gasoline taxes, subsidies for the production of fuel-efficient vehicles, and enhanced research and development programs. The present committee had a much narrower charge. It was directed by Congress, acting through the Department of Transportation (DOT), to concentrate on the impact and effectiveness of Corporate Average Fuel Economy (CAFE) standards originally mandated in the Energy Policy and Conservation Act of 1975. These standards (which have been set at various times both by Congress and by the National Highway Traffic Safety Administration [NHTSA]) establish mandatory fuel efficiencies—in the form of required miles-per-gallon (mpg) goals—for fleets of passenger cars and light-duty trucks, which included the popular sport utility vehicles (SUVs) beginning with the model year (MY) 1978.[1]

It is fair to say that the CAFE program has been controversial since its inception. There are sharp disagreements about the effects of the program on the fuel efficiency of the U.S. vehicle fleet, the current mix of vehicles in that fleet, the overall safety of passenger vehicles, the health of the domestic automobile industry, employment in the industry, and the well-being of consumers. It is this set of concerns and other things that the present committee was asked to address.

But why care about fuel economy at all? It is essential to be clear about the motives for any effort to boost the fuel economy of the vehicle fleet. It is tempting to say that we should care about fuel economy because vehicles that have higher fuel economy ratings (in mpg) will save their owners money. For instance, a car that gets 25 mpg and is driven 15,000 miles/year uses 600 gallons/year of gasoline. Boosting the fuel economy of that car by 20 percent (from 25 to 30 mpg) would save 100 gallons annually—$150 if the price of gasoline is assumed to be $1.50/gallon.[2] The undiscounted savings over an assumed 10-year lifetime would be $1,500. Note that another 5 mpg improvement in fuel economy—from 30 to 35 mpg—would save only about 70 gallons annually, or slightly more than $100. Increases in fuel economy show sharply diminishing returns, an important point in the fuel economy debate.

Yet a strong argument can be made that while these savings are economically relevant, they are not a sufficient basis for public policy intervention. First, consumers already have a wide variety of opportunities if they are interested in better gas mileage. There are many makes and models, all readily available, that get much better than average fuel economy. Second, the differences in fuel economy between vehicles are relatively clear to new car buyers. The Environmental Protection Agency (EPA)-rated fuel economy for each vehicle, in both highway and city driving, is prominently displayed on a sticker on the side of each new car sold. While there are some discrepancies between the EPA fuel economy ratings and buyers' actual experience in on-the-road driving, that constitutes a rationale not for requiring the new vehicle fleet to get better gas mileage but for requiring more accurate fuel economy information on the stickers.

Taking these two points together, it is easy to see that while improved fuel economy saves consumers money, they are quite likely to be both aware of this fact and in a good position to exercise their preference for a more fuel-efficient vehicle if fuel efficiency is an important attribute to them. According to this logic, there is no good reason for the government to intervene and require new cars to get better gas mileage, or to take other policy measures designed to improve the fuel economy of the fleet.

There are, however, other reasons for the nation to consider policy interventions of some sort to boost fuel economy. The most important of these, the committee believes, is concern about the accumulation in the atmosphere of so-called greenhouse gases, principally carbon dioxide (CO_2) (IPCC, 2001; NRC, 2001). Cars and light-duty trucks in the

---

[1] The Corporate Average Fuel Economy program is designed to improve the efficiency of the light-duty vehicle fleet, both automobiles and trucks. It requires vehicle manufacturers to meet a standard in miles per gallon (mpg) for the fleet they produce each year. The standard for automobiles is 27.5 mpg, and for light trucks it is 20.7 mpg. Companies are fined if their fleet average is below the CAFE standard, but various provisions allow flexibility, such as averaging with past and expected fleet averages. Imported and domestic automobile fleets must meet the same standards but are counted separately (trucks are not). The program is administered by the National Highway Traffic Safety Administration (NHTSA) of the Department of Transportation.

Testing is done by manufacturers and spot checked by the Environmental Protection Agency. Vehicles are tested on a dynamometer in a laboratory (to eliminate weather and road variables). Both city and highway driving are simulated and the results combined to compare with the standard. Further information can be found at <http://www.nhtsa.dot.gov/cars/problems/studies/fuelecon/index.html> and <http://www.epa.gov/otaq/mpg.htm>.

[2] Note how small these savings are in relation to the other costs of operating a car each year, such as insurance, vehicle registration fees, parking, and, in a number of states, significant personal property taxes. This point is addressed later in this report.

United States account for slightly less than 20 percent of annual U.S. emissions of CO_2; since the United States accounts for about 25 percent of annual global emissions, these vehicles are responsible for about 5 percent of worldwide annual emissions. Thus, improving the fuel economy of the passenger-car and light-duty-truck fleet would have a nontrivial impact on global CO_2 emissions.[3]

These concerns have been heightened by a series of reports from the Intergovernmental Panel on Climate Change (IPCC), a collection of the world's leading climate scientists. The most recent report suggests that (1) atmospheric concentrations of CO_2 and other greenhouse gases are continuing to increase, (2) the average surface temperature of Earth has increased significantly in the last 100 years, (3) a causal relationship probably exists between 1 and 2, and (4) continued increases in CO_2 emissions could lead to global warming, which would have serious adverse consequences for both plant and animal life on Earth. While also emphasizing the great uncertainties pertaining to climate change, a recent report by the NRC confirmed that "the conclusion of the IPCC that the global warming that has occurred in the last 50 years is likely the result of increases in greenhouse gases accurately reflects the current thinking of the scientific community" (NRC, 2001).

Concerns about climate change are not normally reflected in the market for new vehicles. In this market, the costs that consumers can be expected to take into account are those they will bear directly, including the purchase price of the car and its likely repair costs and resale value over time, expected fuel costs, insurance, taxes, registration, and other costs. Few consumers take into account the environmental costs that the use of their vehicles may occasion. In the parlance of economics, this is a classic negative externality, and it is to be expected that too little fuel economy would be purchased in this case. For that reason, it is appropriate for the government to consider measures that would better align the signals that consumers face with the true costs to society of their use of vehicles. These measures could be of many types—from simple taxes on gasoline designed to internalize the externality to regulatory requirements designed to improve the fuel economy of the vehicles people buy.

There is another reason for concern about automotive fuel economy. In the wake of the oil supply interruptions of the middle and late 1970s, petroleum imports fell as a share of petroleum use. During the 1990s, however, this trend began to reverse. By 2000, imports hit an all-time high of 56 percent of petroleum use and continue to rise. If the petroleum exporters reinvested in the United States all the dollars paid to them, it would not ipso facto be a bad thing. In fact, significant reinvestment has occurred. But such reinvestment may not always be the case, and excessively high levels of imports can put downward pressure on the strength of the dollar (which would drive up the cost of goods that Americans import) and, possibly, increase U.S. vulnerability to macroeconomic instability that can cost the economy considerable real output.

Some experts argue that these vulnerabilities are another form of externality—that is, they are an effect that car buyers do not factor into their decisions but that can represent a true and significant cost to society (see Greene and Tishchishyna, 2000). These experts believe that this justifies government intervention of some sort. Others take a more skeptical view, arguing instead that the macroeconomic difficulties of the 1970s (high unemployment coupled with very high inflation and interest rates) were due more to unenlightened monetary policy than to the inherent difficulties associated with high oil prices (Bohi, 1989). Either way, no one can deny that reducing our nation's oil import bill would have favorable effects on the terms of trade, and that this is a valid consideration in deliberations about fuel economy.

This committee believes it is critically important to be clear about the reasons for considering improved fuel economy. Moreover, and to the greatest extent possible, it is useful to try to think about how much it is worth to society in dollar terms to reduce emissions of greenhouse gases (by 1 ton, say) and reduce dependence on imported oil (by 1 barrel, say). If it is possible to assign dollar values to these favorable effects (no mean feat, the committee acknowledges), it becomes possible to make at least crude comparisons between the beneficial effects of measures to improve fuel economy on the one hand, and the costs (both out-of-pocket and more subtle) on the other.

Having explained why fuel economy matters, why did Congress request a study and why did the NRC create a new committee to examine the issue of CAFE standards? After all, not 10 years have passed since the issuance of the 1992 NRC report, and the current committee believes strongly that the 1992 NRC report is still an excellent place to begin for anyone interested in the fuel economy issue.

In fact, a number of things have changed since the 1992 NRC report that make a reexamination both timely and valuable. Although each of these changes is examined in somewhat more detail in subsequent chapters, it is worth touching on the most important ones here.

First, there have been significant changes in the automobile industry in the last decade. The committee believes it is now virtually meaningless to speak of a U.S. auto company or a Japanese auto company. Today there is a handful of very large companies that both manufacture and sell vehicles

---

[3]At first blush, it might also appear to be the case that fuel economy is important because of the more common air pollutants for which vehicles are responsible—for instance, the precursors for smog (hydrocarbons and oxides of nitrogen), particulate matter, and carbon monoxide. However, automakers are required to meet emissions standards for these pollutants that are denominated in terms of grams per mile traveled. Thus, it should make no difference for emissions of these pollutants whether a car has burned 1 gallon to go 20 miles or 10 gallons—the emissions must be the same. The committee notes, however, that a number of the vehicles achieving high fuel economy also have emission rates that are well below the statutory limits.

around the world. For instance, the General Motors Corporation (GM) has acquired all of Saab and Hummer, half of Isuzu, and a minority share of Suzuki and Fuji Heavy Industries (the makers of Subaru). GM has acquired 20 percent of Fiat Auto S.p.A. (which includes Alfa Romeo), with a full takeover possible, and GM is also in negotiations to acquire Daewoo Motors. For its part, the Ford Motor Company (Ford) has acquired all of Volvo, Jaguar, Aston Martin, and Land Rover, along with a 35 percent stake in Mazda. DaimlerChrysler (itself the product of the largest merger in the history of the auto industry and one of the largest corporate mergers of any type ever) now owns 30 percent of Mitsubishi and 10 percent of Hyundai (Kia). Volkswagen owns all of Rolls-Royce, Bugatti, and Skoda, while Renault has a controlling interest in Nissan as well as Nissan Diesel, and Renault/Nissan has taken over Samsung. Ford and GM have equity stakes in Russian and eastern European assemblers as well. In other words, the auto industry is much, much more concentrated and more global in 2001 than it was in 1991, when the previous committee began its work.

Perhaps equally important, Toyota, Honda, and Nissan—"foreign car makers" whose imports to the United States were a great source of concern a decade ago—all have established significant manufacturing facilities in the United States. This has blurred the distinction between domestic and imported cars, a distinction that was important at the time the original CAFE standards were put in place, and is relevant to current deliberations about the future of the CAFE program. Further blurring any distinctions that used to make sense about imports vs. domestic cars is the North American Free Trade Agreement (NAFTA).

Related to these changes in market structure has been a rather significant change in the financial well-being of the U.S. auto industry and the employment prospects of those who work in it. The United States has always been the largest and most consistently profitable vehicle market in the world, attracting most vehicle producers to sell vehicles to it. In recent years, the U.S. market has been strengthened by, among other things, the scale economies resulting from the mergers discussed above; the remarkable performance of the U.S. economy during the 1990s; and the exploding popularity of minivans, pickup trucks, and sport utility vehicles (SUVs). Automakers headquartered in the United States regained their financial health and, with the exception of the Chrysler component of DaimlerChrysler, the balance sheets of the carmakers were in good shape.[4] During the 1990s, Ford, GM, and Chrysler bought back billions of dollars of their own shares with the excess cash they generated and invested heavily in other businesses, ranging from financial services to car repair and salvage. At the same time, employment in the U.S. auto industry reached a new peak as foreign and domestic manufacturers expanded capacity at existing plants, and foreign manufacturers invested in new auto assembly and parts plants.

Another significant change over the last decade has to do with the trend in automobile fuel economy. At the time of the 1992 report, the average fuel economy of the entire U.S. light-vehicle fleet had just begun to decrease after nearly 15 years of improvements totaling more than 66 percent (see Chapter 2). By 2000, however, overall light-vehicle fuel economy had not only failed to reverse the trend seen between 1988 and 1991 but had fallen still farther—it is now about 7 percent lower than at its peak in 1987–1988 (EPA, 2000).

This change, in turn, is almost a direct consequence of another dramatic change in the automobile market, one familiar to each and every driver and passenger—the shift in the mix of vehicles on the road away from traditional passenger cars and toward SUVs, pickup trucks, and vans (which, collectively, are referred to as light-duty trucks). As recently as 1975 (the year the CAFE program was legislated into existence), traditional passenger cars accounted for fully 80 percent of the light vehicle market. By 2000, the light-duty truck component of the light vehicle market had grown to 46 percent, and its share is expected to exceed that of passenger cars for the first time in 2001. Because new light-duty trucks sold are required to meet a fleet average fuel economy standard of 20.7 mpg compared with the 27.5 mpg standard that currently applies to the passenger car fleet, this shift has been pulling down the overall fuel economy of the light vehicle fleet.

Moreover, a new type of vehicle has appeared on the market and is growing in popularity. This is the crossover vehicle (a light truck), which has the appearance of an SUV and many of its characteristics but which is built on a passenger car platform rather than a light truck platform (examples are the Lexus RX-300 and the Toyota RAV-4). It is too soon to predict whether these vehicles will penetrate deeply into the light-duty vehicle market and what their effect would be if they do (for instance, will they replace station wagons and large cars, worsening overall fuel economy, or will they become smaller and more fuel-efficient substitutes for larger SUVs?). It is a change that bears close watching.

Another development since 1992 has to do with new technologies for vehicular propulsion. For instance, both Toyota (the Prius) and Honda (the Insight) have already introduced into the vehicle market the first hybrid-electric passenger cars—vehicles that combine a traditional internal combustion engine powered by gasoline with an electric motor that assists the engine during acceleration. These cars recapture some of the energy lost during braking and can shut off their engines instead of idling, with almost instant restart, both of which are important energy-saving features. The Insight, which seats two passengers and weighs about 2,000 lb, has an EPA rating of 61 mpg in city driving and 70 mpg in highway driving. The Prius gets 52 mpg in the city and 45 mpg

---

[4]This will not be the case if the U.S. economy enters a protracted downturn and/or if profits on light-duty trucks narrow significantly.

on the highway. Ford has announced it will begin selling a hybrid SUV, the Escape, in 2004 and has pledged to boost the fuel economy of its SUVs by 25 percent by 2004. GM and DaimlerChrysler have pledged to outdo any improvements Ford makes on SUV fuel economy.

At the same time, there is great excitement and a steady stream of progress reports about the fuel cell. Fuel cells hold promise for alleviating the problems associated with fossil fuel combustion in both stationary (e.g., electric power plant) and mobile (e.g., vehicle power plant) sources. This is because they produce power without the combustion processes that generate conventional air pollutants such as particulate matter, CO_2, and other undesirable by-products (see Chapter 4, Attachment 4A, for a discussion of full fuel-cycle impacts). Moreover, the fuel cell and other alternative technologies have been given a boost by the government/industry cooperative venture Partnership for a New Generation of Vehicles (PNGV), aimed at the development of a midsize automobile that is safe, affordable, and capable of getting dramatically better fuel economy (up to 80 mpg) (NRC, 2000).

Technologies have changed in other respects as well since the earlier 1992 NRC report. During the 1990s, automakers improved the performance characteristics of their light-duty vehicles considerably. For instance, the horsepower-to-weight ratio for passenger cars and light trucks is up about 50 percent since 1981. Similarly, the time it takes for a vehicle to accelerate from 0 to 60 miles per hour has fallen 26 percent since 1981—and 2 percent in the last year alone. Another way of saying this is that automakers have indeed made considerable technological advances in the cars and light trucks they made and sold during the last decade. But these advances have almost all been aimed at making cars faster and more powerful, at selling more and heavier light trucks, and at equipping vehicles with other extras (heated seats, power windows, and cruise control, for example) rather than at making them more fuel efficient. This is perfectly understandable, incidentally, given the apparent lack of interest in fuel economy on the part of the car-buying public at mid-2001 gasoline prices.

Since the 1992 NRC report, moreover, EPA has issued new Tier 2 emissions standards under the 1990 amendments to the Clean Air Act. These standards affect not only emissions but also certain technologies such as the advanced diesel engine that could be used to improve overall fuel economy.

A final factor suggests a fresh look at fuel economy and the way in which it has been and could be affected by the CAFE program. Specifically, over the last 20 years—and perhaps particularly over the last decade—there has been a steady increase in the attention that car buyers pay to safety concerns. This is one of the causes of the steady, long-term decline in the fatality rate per vehicle mile traveled. Whatever the reason, safety "sells" in a way that was almost inconceivable two decades ago. This is germane to the committee's work because the possible effects on safety of the original CAFE program, as well as the effects on safety that a renewed effort to improve fuel economy would have, have been perhaps the most controversial aspect of the program. Because we now have another decade's worth of research on the determinants of vehicle safety, a fresh look at automotive fuel economy is warranted.

SCOPE AND CONDUCT OF THE STUDY

In legislation for fiscal year 2001, Congress requested that the National Academy of Sciences, in consultation with DOT, conduct a study to evaluate the effectiveness and impacts of CAFE standards.[5] In particular, it asked that the study examine the following, among other factors:

1. The statutory criteria (economic practicability, technological feasibility, need for the United States to conserve energy, the classification definitions used to distinguish passenger cars from light trucks, and the effect of other regulations);
2. The impact of CAFE standards on motor vehicle safety;
3. Disparate impacts on the U.S. automotive sector;
4. The effect on U.S. employment in the automotive sector;
5. The effect on the automotive consumer; and
6. The effect of requiring separate CAFE calculations for domestic and nondomestic fleets.

In consultation with DOT, a statement of work for the committee was developed (see Appendix D). The committee's work was to emphasize recent experience with CAFE standards, the impact of possible changes, and the stringency and/or structure of the CAFE program in future years.

The committee conducted numerous meetings and made several site visits during the short time frame of this study. It held open sessions during several of its meetings to receive presentations from a wide variety of individual experts and representatives of the private sector, nongovernmental organizations, environmental groups, and government and to collect information and data on the various issues related to CAFE standards. Also, many reports, statements, and analyses were submitted to the committee for its review. The committee also used consultants under its direction to facilitate its work under the tight time constraints of the study. For example, Energy and Environmental Analysis, Inc. (EEA) conducted analyses of potential improvements in fuel economy and related costs for a number of different vehicle classes. Sierra Research provided insight to the committee

---

[5]Conference Report on H.R. 4475, Department of Transportation and Related Agencies Appropriations Act, 2001. Report 106-940, as published in the *Congressional Record,* October 5, 2000, pp. H8892-H9004.

on fuel economy improvements and costs, based on work it had done for the automotive companies. In the end, however, the committee conducted its own analyses, informed by the work of the consultants, the technical literature, presentations at its meetings, material submitted to it, and the expertise and judgment of the committee members, to arrive at its own range of estimates for fuel economy improvements and associated costs.

In conducting its study, the committee first assessed the impact of the current CAFE system on reductions in fuel consumption, on greenhouse gases, on safety, and on impacts on the industry (see Chapter 2). To assess what the impacts of changed fuel economy standards might be, it examined opportunities for fuel efficiency improvements for vehicles with the use of existing or emerging technologies, estimated the costs of such improvements, and examined the lead times that would be required to introduce the vehicle changes (see Chapter 3). Based on these examinations, the implications for changed CAFE standards are presented in Chapter 4. The committee also examined the stringency and structure of the CAFE system and assessed possible modifications of the system, as well as alternative approaches to achieving greater fuel economy for vehicles, which resulted in suggestions for improved policy instruments (see Chapter 5). Chapter 6 contains the committee's findings and recommendations. Appendix E is a list of acronyms and abbreviations.

Following the release of the prepublication copy of the report in July 2001, the committee reexamined its technical and economic analysis, as discussed in Appendix F. Minor changes have been made to some of the material in Chapters 3 and 4 as a result of this reexamination, but the findings and conclusions are substantively unchanged.

REFERENCES

Bohi, D. 1989. Energy Price Shocks and Macroeconomic Performance. Washington, D.C.: Resources for the Future.

Environmental Protection Agency (EPA). 2000. Light-Duty Automotive Technology and Fuel Economy Trends 1975 Through 2000. EPA420-R00-008 (December). Office of Air and Radiation. Washington, D.C.: Environmental Protection Agency.

Greene, D., and N.I. Tishchishyna. 2000. Costs of Oil Dependence: A 2000 Update (May). ORNL/TM-2000/152. Oak Ridge, Tenn.: Oak Ridge National Laboratory.

Intergovernmental Panel on Climate Change (IPCC). 2001. Climate Change 2001: The Scientific Basis. Cambridge, U.K.: Cambridge University Press.

National Research Council (NRC). 1992. Automotive Fuel Economy: How Far Should We Go? Washington, D.C.: National Academy Press.

NRC. 2000. Review of the Research Program of the Partnership for a New Generation of Vehicles, Sixth Report. Washington, D.C.: National Academy Press.

NRC. 2001. Climate Change Science: An Analysis of Some Key Questions. Washington, D.C.: National Academy Press.

2

The CAFE Standards: An Assessment

Twenty-five years after Congress enacted the Corporate Average Fuel Economy (CAFE) standards, petroleum use in light-duty vehicles is at an all-time high. It is appropriate to ask now what CAFE has accomplished, and at what cost. This chapter begins by addressing energy and CAFE: What is the current rationale for fuel economy standards? How have vehicles changed, in particular in regard to fuel economy? What is the impact on oil consumption? The first section addresses a series of questions the committee was asked about the impact of CAFE. The second section explores the impact of CAFE on the automotive industry. The final section reviews the impact on safety.

Isolating the effects of CAFE from other factors affecting U.S. light-duty vehicles over the past 25 years is a difficult analytical task. While several studies have tried to estimate the specific impacts of CAFE on fuel economy levels and on highway safety, there is no comprehensive assessment of what would have happened had fuel economy standards not been in effect. Lacking a suitable baseline against which to compare what actually did happen, the committee was frequently unable to separate and quantify the impacts of factors such as fuel prices or of policies such as the gas guzzler tax. Much of this report describes what happened before and after the implementation of the CAFE standards, with little or no isolation of their effects from those of other forces affecting passenger cars and light trucks. While this analytical approach is less than ideal, it can provide some sense of whether the impacts were large or small, positive or negative.

CAFE AND ENERGY

Rationale for Fuel Economy Standards

The nation's dependence on petroleum continues to be an economic and strategic concern. Just as the 1992 committee (the fuel economy committee) cited the conflict in the Persian Gulf as evidence of the fragility of the world's petroleum supply, the current committee cites the oil price hikes of 1999 and 2000 as further evidence of the nation's need to address the problem of oil dependence. The association between oil price shocks and downturns in the U.S. economy (see Figure 2-1) has been documented by numerous studies over the past 20 years (for example, Hamilton, 1983 and 1996; Hickman, 1987; Huntington, 1996; Mork et al., 1994). While the causes of recessions are complex and other factors, such as monetary policy, play important roles, oil price shocks have clearly been a contributing factor (Darby, 1982; Eastwood, 1992; Tatom, 1993). Estimates of the cumulative costs to our economy of oil price shocks and noncompetitive oil pricing over the past 30 years are in the trillions of dollars (see, for example, Greene and Tishchishyna, 2000; EIA, 2000c; DOE, 1991; Greene et al., 1998).

Today, oil is a much less important share of the economy (expenditures on oil amount to 2 percent of the gross domestic product [GDP]) than it was in the early 1980s but approximately the same as in 1973, the year of the first Arab-OPEC (Organization of Petroleum Exporting Countries) oil embargo. Still, U.S. oil imports now exceed 50 percent of consumption and are projected to increase substantially (EIA, 2000a, table A.11). The U.S. transportation sector remains nearly totally dependent on petroleum, and passenger cars and light trucks continue to account for over 60 percent of transportation energy use (Davis, 2000, table 2.6).

A second petroleum-related factor, possibly even more important, has emerged since CAFE was enacted: global climate change. Scientific evidence continues to accumulate supporting the assertion that emissions of greenhouse gases from the combustion of fossil fuels are changing Earth's climate.

International concern over the growing emissions of greenhouse gases from human activities has increased substantially since the 1992 assessment of fuel economy by the National Research Council (NRC, 1992, pp. 70–71). The scientific evidence suggesting that emissions of CO_2 and other greenhouse gases are producing global warming, causing the

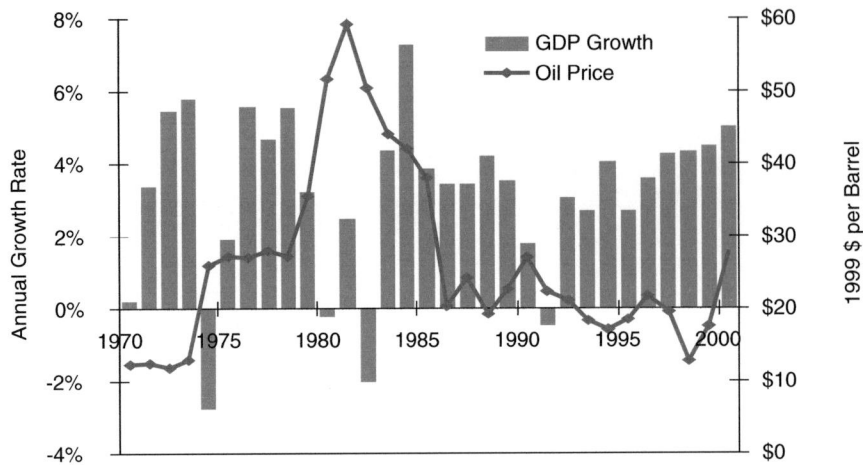

FIGURE 2-1 Oil price shocks and economic growth, 1970–1999. SOURCE: Adapted from Greene and Tishchishyna (2000).

sea level to rise, and increasing the frequency of extreme weather events has grown stronger (IPCC, 2001, pp. 1–17; NRC, 2001). Concern over the potentially negative consequences of global climate change has motivated the European Union and Japan to take steps to reduce CO_2 emissions from passenger cars and light trucks by adopting new fuel economy standards (Plotkin, 2001).

The transportation sector accounts for about 31 percent of anthropogenic CO_2 emissions in the U.S. economy; CO_2 accounts for over 80 percent of greenhouse gas emissions from the economy as a whole (EIA, 2000b). Since the United States produces about 25 percent of the world's greenhouse gases, fuel economy improvements could have a significant impact on the rate of CO_2 accumulation in the atmosphere. However, it should be noted that other sectors, particularly electricity, have far more potential for reducing CO_2 emissions economically (EIA, 1998). Focusing on transportation alone would accomplish little.

New Car and Light Truck Fuel Economy

The CAFE standards, together with significant fuel price increases from 1970 to 1982, led to a near doubling of the fuel economy of new passenger cars and a 50 percent increase for new light trucks (NRC, 1992, p. 169) (see Figure 2-2). While attempts have been made to estimate the relative contributions of fuel prices and the CAFE standards to this improvement (see, for example, Crandall et al., 1986; Leone and Parkinson, 1990; Greene, 1990; Nivola and Crandall, 1995), the committee does not believe that responsibility can be definitively allocated. Clearly, both were important, as were efforts by carmakers to take weight out of cars as a

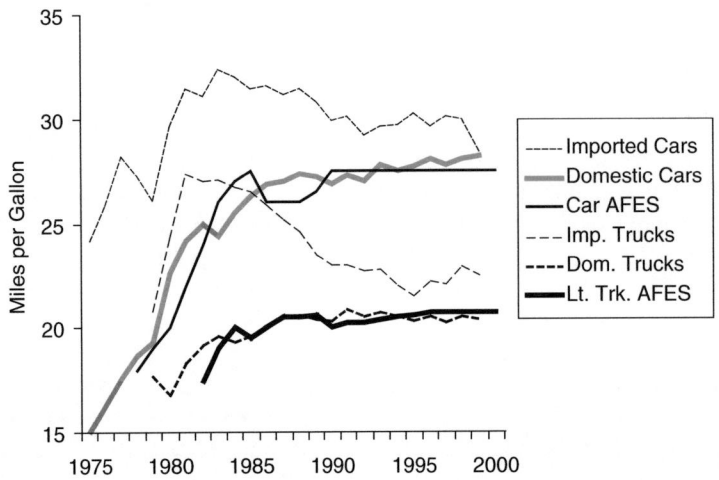

FIGURE 2-2 Automotive fuel economy standards (AFES) and manufacturers' CAFE levels. SOURCE: Based on NRC (1992) and EPA (2000).

cost-saving measure. CAFE standards have played a leading role in preventing fuel economy levels from dropping as fuel prices declined in the 1990s.

The increasing market share of higher fuel economy imported vehicles was also a factor in raising the average fuel economy of the U.S. fleet and decreasing its average size and weight. The market share of foreign-designed vehicles increased from 18 percent in 1975 to 29 percent in 1980 and 41 percent in 2000. In 1975, the average foreign-designed vehicle achieved about 50 percent higher fuel economy than the average domestic vehicle. Foreign-designed vehicles also weighed about 40 percent (1,700 lb) less (EPA, 2000, tables 14 and 15). These differences have narrowed considerably over time, as shown below.

Figure 2-2 suggests that the CAFE standards were not generally a constraint for imported vehicles, at least until 1995, if then. Domestic manufacturers, on the other hand, made substantial fuel economy gains in line with what was required by the CAFE standards. The fuel economy numbers for new domestic passenger cars and light trucks over the past 25 years closely follow the standards. For foreign manufacturers, the standards appear to have served more as a floor toward which their fuel economy descended in the 1990s.

For the most part, the differing impacts of the CAFE standards on domestic and foreign manufacturers were due to the different types of vehicles they sold, with foreign manufacturers generally selling much smaller vehicles than domestic manufacturers.

In 1975, when CAFE was enacted, 46 percent of the cars sold by domestic manufacturers were compacts or smaller, while 95 percent of European imports and 100 percent of Asian imports were small cars.

The difference between the product mix of domestic manufacturers and that of foreign manufacturers has dramatically narrowed since then. By 2000, small cars represented 39 percent of the market for domestic carmakers (down from 46 percent in 1975) and had plummeted to 60 percent and 50 percent for European and Asian manufacturers, respectively, based on interior volume (EPA, 2000, appendix K). This convergence is also evident when the average weights of domestic and imported vehicles in given market segments are compared (see Figure 2-3). In 1975, the average weight of a domestic passenger car was 4,380 lb. It outweighed its European counterpart by 1,676 lb and its Asian counterpart by 1,805 lb. In 2000, the average domestic passenger car weighed 75 lb less than the average European car and only 245 lb more than the average Asian passenger car. What had been a 70 percent difference between the average weights of domestic and Asian cars decreased to 7.6 percent. There is now little difference in the market positions of domestic and imported manufacturers, as a whole, in the passenger car market.

Another factor contributing to the superior fuel economy of imported automobiles in 1975 was technology. Only 1.3 percent of domestic passenger cars used front-wheel drive in 1975, compared with 17 percent of Asian imports and 46 percent of European imports. Similarly, less than 1 percent of domestic cars were equipped with fuel injection that year, while 14 percent of Asian imports and 39 percent of European imports used that more efficient technology. Undoubtedly, higher fuel prices in Europe and Asia were (and still are) a major incentive for rapid implementation of fuel economy technologies. However, the emphasis on small cars in 1975 by foreign manufacturers was clearly the most important reason for their higher fuel economy.

The light-truck market has fared differently. While the weights of vans have converged somewhat, domestic pickups are still about 13 percent heavier than their imported counterparts. Although the difference in average weight between domestic and imported sport utility vehicles (SUVs)

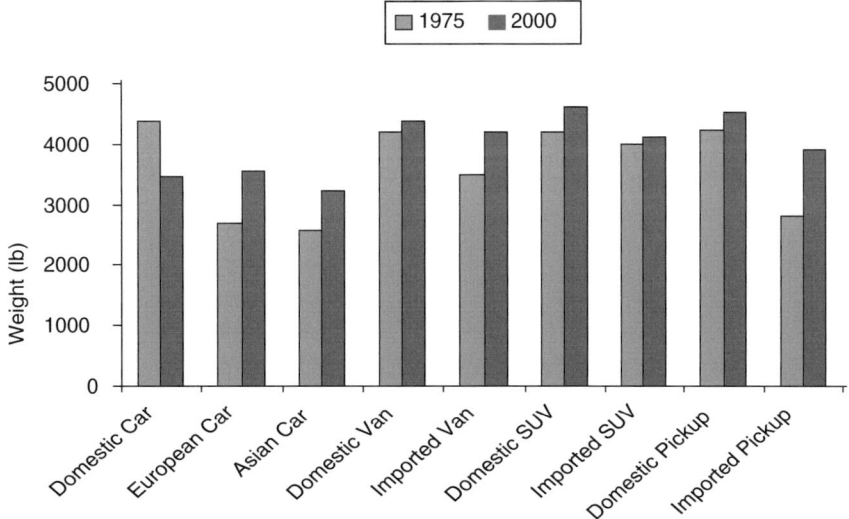

FIGURE 2-3 Average weights of domestic and imported vehicles. SOURCE: EPA (2000).

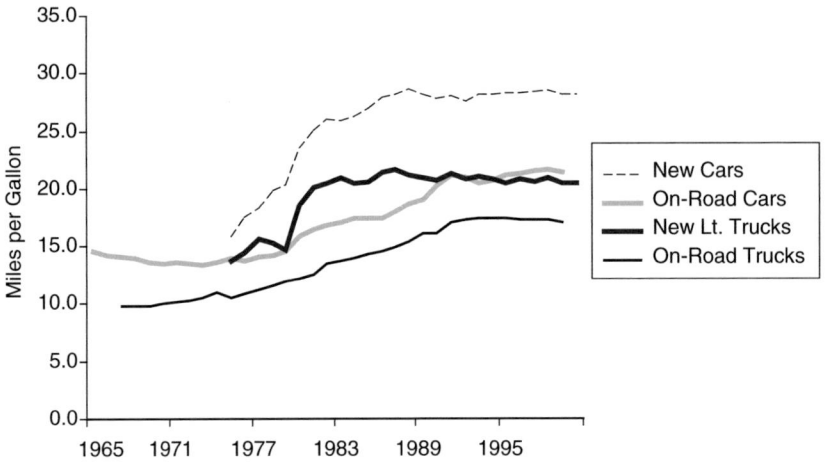

FIGURE 2-4 Fleet fuel economy of new and on-road passenger cars and light trucks. SOURCE: FHWA (2000).

seems to have increased, the SUV market, as it is today, did not exist in 1975. Similar differences exist by size class. Only 1.3 percent of domestic light trucks are classified as small, 44 percent as large. By contrast, 36 percent of imported trucks are small and only 6 percent are large (EPA, 2000). Improvements in new vehicle fuel economy have gradually raised the overall fuel economy of the entire operating fleet as new vehicles replace older, less fuel-efficient vehicles.[1] DOT's Federal Highway Administration (FHWA) estimates that the average miles per gallon (mpg) for all passenger cars in use—both old and new—increased from 13.9 mpg in 1975 to 21.4 mpg in 1999 (see Figure 2-4) (FHWA, 2000). The estimated on-road fuel economy of light trucks improved from 10.5 to 17.1 mpg over the same period. Since the FHWA's definitions of passenger car and light truck are not the same as those used for CAFE purposes, it is also useful to consider the trend in combined light-duty vehicle fuel economy. The FHWA estimates that overall light-duty vehicle on-road fuel economy increased from 13.2 mpg in 1975 to 19.6 mpg in 1999, a gain of 48 percent. The EPA sales-weighted test numbers indicate that new light-duty vehicle fuel economy increased from 15.3 mpg in 1975 to 24 mpg in 1999, a gain of 57 percent. Given the remaining older vehicle stock yet to be retired and the inclusion of some larger two-axle, four-tire trucks in the FHWA's definition, these numbers are roughly comparable.

As Figure 2-4 illustrates, there is a substantial shortfall between fuel economy as measured for CAFE purposes and actual fuel economy achieved on the road. If the EPA ratings accurately reflected new vehicle fuel economy, the operating fleet averages would be approaching those levels. Instead, they are leveling off well below the ratings. The shortfall, which the EPA estimates at about 15 percent, is the result of a number of factors that differ between actual operating conditions and the EPA test cycle, such as speed, acceleration rates, use of air conditioners, and trip lengths (Hellman and Murrell, 1984; Harrison, 1996). A comparison of FHWA on-road and EPA new vehicle fuel economy estimates suggests a larger discrepancy for passenger cars than for light trucks. This pattern is contrary to the findings of Mintz et al. (1993), who found a larger shortfall for light trucks of 1978–1985 vintages. The discrepancy may reflect a combination of estimation errors and differences in definitions of the vehicle types.[2] The EPA estimates that new light-duty vehicles have averaged 24 to 25 mpg since 1981. The FHWA estimates the on-road fuel economy of all light-duty vehicles at 19.6 mpg in 1999, a difference of about 20 percent. Some of this discrepancy reflects the fact that a substantial number of pre-1980 vehicles (with lower fuel economy) were still on the road in 1999, but most probably reflects the shortfall between test and on-road fuel economy.

Vehicle Attributes and Consumer Satisfaction

Significant changes in vehicle attributes related to fuel economy accompanied the fuel economy increases brought about by CAFE standards, fuel price increases, and manufacturers' efforts to reduce production costs. Between 1975

---

[1]The lag is due to the time required to turn over the vehicle fleet. Recent estimates of expected vehicle lifetimes suggest that an average car will last 14 years and an average light truck 15 years (Davis, 2000, tables 6.9 and 6.10). This means that about half of the vehicles sold 15 years ago are still on the road today.

[2]The light truck definition used by FHWA for traffic monitoring differs substantially from that used by the National Highway Traffic Safety Administration (NHTSA) for CAFE purposes. The chief difference is that FHWA's definition includes larger light trucks not covered under the CAFE law. In addition, the FHWA's division of fuel use and vehicle miles traveled (VMT) between passenger cars and light trucks is generally considered to be only approximately correct. It is probably more accurate to compare combined light-duty vehicle mpg estimates.

and 1980, the fuel economy of new passenger cars increased by 50 percent, from 15.1 to 22.6 mpg. At the same time, the size and weight of passenger cars decreased significantly (see Figure 2–5). The average interior volume of a new car shrank from 111 cubic feet in 1975 to 105 cubic feet in 1980. Decreasing interior volume, however, appears to have been part of a trend extending back to the 1960s, at least. Average passenger car wheelbase also declined—from 110 inches in 1977 to 103 in 1980. Curb weight simultaneously decreased by more than 800 lb. This reduction in weight was clearly not part of a previous trend.

From 1980 to 1988, passenger car characteristics changed little, while new vehicle fuel economy improved by 19 percent, from 24.3 mpg in 1980 to an all-time high of 28.8 mpg in 1988. Since then, new vehicle fuel economy has remained essentially constant, while vehicle performance and weight have increased. For passenger cars, horsepower, acceleration (hp/lb), and top speed all continued to increase in line with a trend that began in 1982 (see Figure 2-6). Between 1975 and 1980, in contrast, passenger car weight and horsepower decreased, acceleration and top speed remained nearly constant, and fuel economy increased sharply.

Light truck attributes show similar patterns, although nearly all of the increase in light-truck fuel economy was accomplished in the 2 years between 1979 and 1981 (see Figure 2-7). The weight of light trucks did not decline as sharply as the weight of passenger cars, and in recent years it has reached new highs. Weight, horsepower-to-weight ratios, and top speeds have all been increasing since 1986.

Other engineering and design changes, motivated at least in part by the need to increase fuel economy, probably influenced consumers' satisfaction with new vehicles. For example, front-wheel drive, which affects handling and improves traction, also permits weight reduction owing to the elimination of certain drive train components and repackaging. Use of front-wheel drive in passenger cars increased from 6.5 percent in 1975 to 85 percent by 1993, where it more or less remains today. Less than 20 percent of light

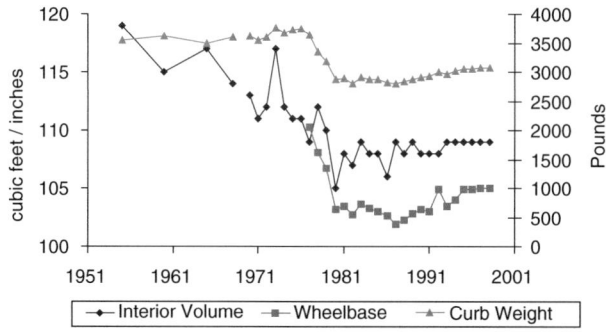

FIGURE 2-5 Passenger car size and weight. SOURCE: Orrin Kee, National Highway Traffic Safety Administration, production-weighted data from manufacturers' fuel economy reports, personal communication.

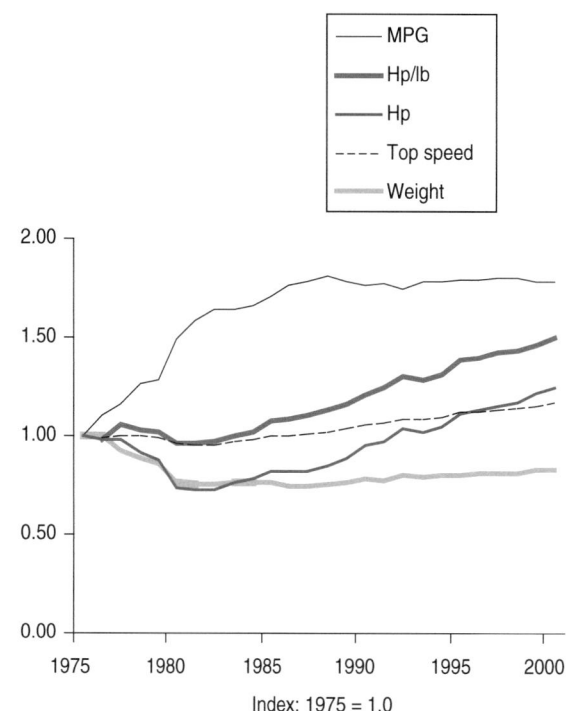

FIGURE 2-6 Trends in fuel-economy-related attributes of passenger cars, 1975–2000. SOURCE: EPA (2000).

trucks employ front-wheel drive, but none did in 1975. Seventy-five percent of vans use front-wheel drive. Fuel injection, which improves fuel metering for more efficient combustion, is also essential for meeting today's pollutant emission standards and improves engine responsiveness as

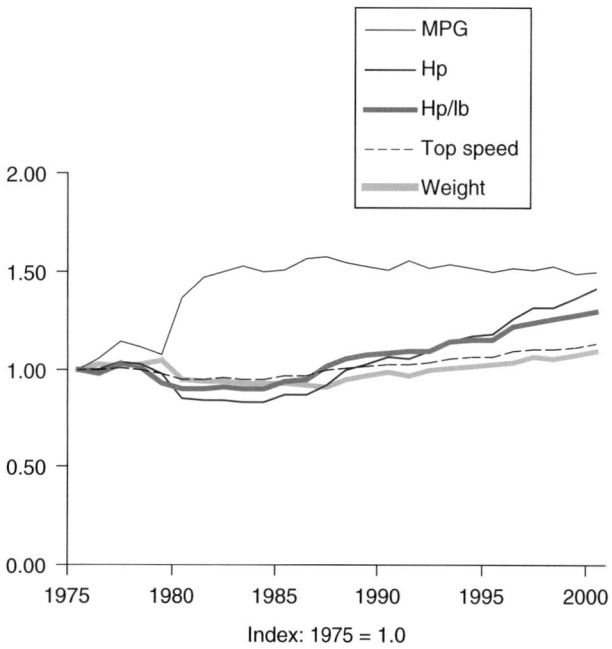

FIGURE 2-7 Trends in fuel-economy-related attributes of light trucks, 1975–2000. SOURCE: EPA (2000).

well. Use of fuel injection increased from 5 percent for passenger cars and 0 percent for light trucks in 1975 to 100 percent for both categories today. Use of lock-up torque converters in automatic transmissions, which reduce slip and thereby increase efficiency, increased from 0 percent in 1975 to 85 percent today for both passenger cars and light trucks. Use of four-valve-per-cylinder engines increased from 0 percent before 1985 to 60 percent in passenger cars, 20 percent in vans, 25 percent in SUVs, and only 1 percent in pickup trucks. Four-valve engines offer improved fuel economy and performance over a wide range of speeds.

Fuel economy improvements have affected the costs of automobile ownership and operation over the past 25 years. However, the precise impacts on vehicle price and customer satisfaction are not known because of the lack of accurate accounting of the costs of fuel economy improvements and the difficulty of attributing changes in vehicle attributes to fuel economy or to other design goals. How to attribute the costs of numerous technology and design changes to the CAFE standards or to other factors, such as fuel prices, is unclear. Nonetheless, it is possible to examine trends in the overall costs of owning and operating passenger cars, which shed some light on the possible impacts of the CAFE standards.

The cost (in constant dollars) of owning and operating automobiles appears to be only slightly higher today than in 1975. The American Automobile Association estimates that the total cost per mile of automobile ownership in 1975 was 55.5 cents, in constant 1998 dollars (Davis, 2000, table 5.12). The estimate for 1999 was 56.7 cents per mile.[3] Fixed costs (costs associated with owning or leasing a vehicle that are not directly dependent on the miles driven), which today account for more than 80 percent of total costs, were at least 30 percent higher in 1999 than in 1975. Operating costs have been nearly halved, mostly because of the reduction in gas and oil expenditures. In 1975, expenditures on gas and oil were estimated to be 14.6 cents per mile and constituted 75 percent of variable (or operating) costs (26 percent of total costs). In 1999, gas and oil expenses were only 5.5 cents per mile, just over 50 percent of variable costs and less than 10 percent of total costs. A large part of the change has to do with the lower price of gasoline in 1999, $1.17 per gallon in 1999 versus $1.42 in 1975 (1996 dollars) (Davis, 2000). The rest is the result of improvements in fuel economy.

The average price of a new automobile has increased from just under $15,000 in 1975 to over $20,000 today (1998 dollars). Virtually all of the price increase came after 1980, by which time most of the increase in passenger car fuel economy had already been accomplished (see Figure 2-8). Furthermore, the average purchase price of imported cars, which were largely unconstrained by the CAFE standards

---

[3]These costs are not strictly comparable, however, due to a change in the method of estimating depreciation instituted in 1985. Because of this change, fixed costs prior to 1985 are inflated relative to later costs.

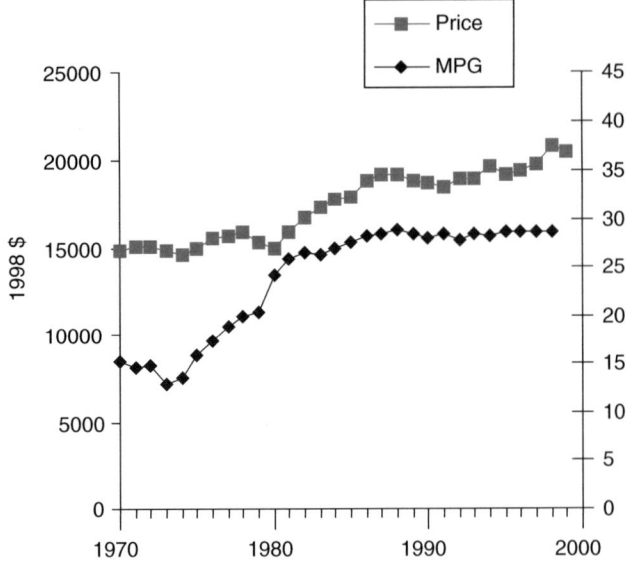

FIGURE 2-8 Average new car price and fuel economy. SOURCE: Based on Davis (2000) and EPA (2000).

(because most of the vehicles sold by foreign manufacturers were above the standard), has increased far more than that of domestic cars, which were constrained. The average price of large trucks has risen faster than that of passenger cars.

The committee heard it said that CAFE may have instigated the shift from automobiles to light trucks by allowing manufacturers to evade the stricter standards on automobiles. It is quite possible that CAFE did play a role in the shift, but the committee was unable to discover any convincing evidence that it was a very important role. The less stringent CAFE standards for trucks did provide incentives for manufacturers to invest in minivans and SUVs and to promote them to consumers in place of large cars and station wagons, but other factors appear at least as important. Domestic manufacturers also found light-truck production to be very attractive because there was no foreign competition in the highest-volume truck categories. By shifting their product development and investment focus to trucks, they created more desirable trucks with more carlike features: quiet, luxurious interiors with leather upholstery, top-of-the-line audio systems, extra rows of seats, and extra doors. With no Japanese competition for large pickup trucks and SUVs, U.S. manufacturers were able to price the vehicles at levels that generated handsome profits. The absence of a gas guzzler tax on trucks and the exemption from CAFE standards for trucks over 8,500 lb also provided incentives.

Consumers also found many of these new vehicles very appealing. They offer roomy interiors that accommodate many passengers, ample storage space, towing capacity, good outward visibility, and a sense of safety and security. Midsize SUVs rose from 4.0 percent of all light-duty vehicle sales in MY 1988 to 12.3 percent in 2000. Midsize station wagons dropped from 1.9 to 1.4 percent over the same period. Large SUVs rose from 0.5 percent to 5.5 percent, while large station wagons dropped from 0.5 percent to zero (EPA, 2000). SUVs are far more popular today than station wagons were before CAFE. Furthermore, several wagons, including the Toyota Camry, Honda Accord, and Nissan Maxima, were dropped from production even though the manufacturers were not constrained by CAFE. Therefore, it must be concluded that the trend toward trucks probably would have happened without CAFE, though perhaps not to the same degree.

The effect of the shift to trucks on fuel economy has been pronounced. As shown in Figures 2-6 and 2-7, the fuel economy of new cars and trucks, considered separately, has been essentially constant for about 15 years. However, the average fuel economy of all new light-duty vehicles slipped, from a peak of 25.9 mpg in 1987 to 24.0 mpg in 2000, as the fraction of trucks increased from 28 to 46 percent (EPA, 2000). Even if trucks and cars maintain their current shares, the average fuel economy of the entire on-road fleet will continue to decline as new vehicles replace older ones with their higher fraction of cars.

Impact on Oil Consumption and the Environment

Fuel use by passenger cars and light trucks is roughly one-third lower today than it would have been had fuel economy not improved since 1975, as shown in this section. As noted above, the CAFE standards were a major reason for the improvement in fuel economy, but other factors, such as fuel prices, also played important roles.

Travel by passenger cars and light trucks has been increasing at a robust average annual rate of 3.0 percent since 1970 (see Figure 2-9). Growth has been relatively steady, with declines in vehicle miles traveled (VMT) occurring only during the oil price shocks and ensuing recessions of 1973–1974, 1979–1980, and 1990–1991. Throughout this period, light-truck travel has been growing much more rapidly than automobile travel. From 1970 to 1985, light-truck VMT grew at an average rate of 7.7 percent/year, while passenger car VMT grew at only 1.8 percent/year. From 1985 to 1999, the rates were similar: 6.1 percent/year for light trucks and 1.7 percent/year for passenger cars. The trend toward light trucks appears to antedate the CAFE standards. From 1966[4] to 1978, light-truck VMT grew 9.8 percent/year, while passenger-car VMT grew at 3.3 percent/year.

Prior to 1978, fuel use by passenger cars and light trucks was growing slightly faster than VMT (see Figure 2-9). It then declined from 1978 to 1982 as gasoline prices soared and the first effects of the higher fuel economy of new vehicles began to have an impact on the fleet (see Figure 2-7). While it is difficult to say what fuel consumption would have been had there been no CAFE standards, it is clear that if light-duty fuel use had continued to grow at the same rate as light-duty VMT, the United States would be currently consuming approximately 55 billion more gallons of gasoline each year (equivalent to about 3.6 million barrels per day [mmbd] of gasoline).

On the other hand, increased fuel economy also reduces the fuel cost per mile of driving and encourages growth in vehicle travel. Estimates of the significance of this "rebound effect" suggest that a 10 percent increase in fuel economy is likely to result in roughly a 1 to 2 percent increase in vehicle travel, all else being equal (Greene et al., 1999; Haughton

---

[4]The FHWA substantially changed its truck class definitions in 1966, making that the earliest date for which there is a consistent definition of a "light truck."

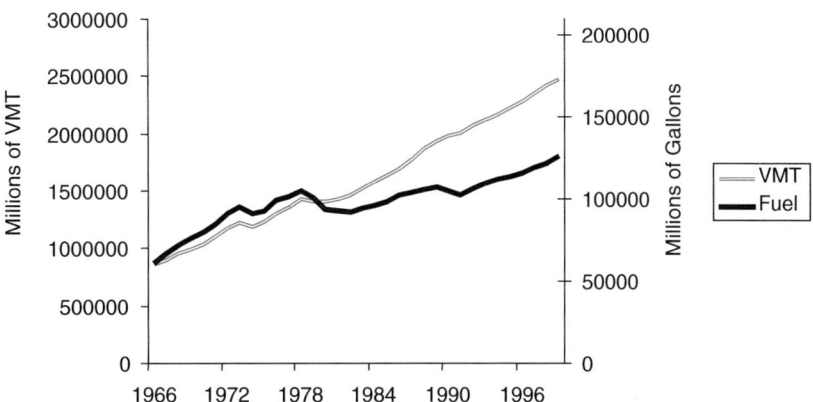

FIGURE 2-9 Passenger car and light-truck travel and fuel use. SOURCE: Based on Davis (2000).

and Sarker, 1996; Jones, 1993). Applying that to the estimated 44 percent increase in on-road, light-duty fuel economy from 1975 to 2000 would reduce the estimated annual fuel savings from 55 billion to 43 billion gallons, equivalent to about 2.8 mmbd of gasoline.

Reducing fuel consumption in vehicles also reduces carbon dioxide emissions. If the nation were using 2.8 mmbd more gasoline, carbon emissions would be more than 100 million metric tons of carbon (mmtc) higher. Thus, improvements in light-duty vehicle fuel economy have reduced overall U.S. emissions by about 7 percent. In 1999, transportation produced 496 mmtc, about one-third of the U.S. total. Passenger cars and light-duty trucks accounted for about 60 percent of the CO_2 emissions from the U.S. transportation sector (EPA, 2001), or about 20 percent of total U.S. emissions of greenhouse gases. Overall, U.S. light-duty vehicles produce about 5 percent of the entire world's greenhouse gases.

Impact on Oil Markets and Oil Dependence

The fuel economy of U.S. passenger cars and light trucks affects world oil markets because U.S. light-duty vehicles alone account for 10 percent of world petroleum consumption. Reducing light-duty vehicle fuel use exerts downward pressure on world oil prices and on U.S. oil imports. Together with major increases in non-OPEC oil supply, reductions in petroleum demand in the United States and other countries created the conditions for the collapse of OPEC market power in 1986. Had past fuel economy improvements not occurred, it is likely that the U.S. economy would have imported more oil and paid higher prices than it did over the past 25 years. CAFE standards have contributed to past light-duty vehicle fuel economy improvements, along with past fuel price increases and other factors.

Oil price shocks have had serious economic consequences for oil-consuming nations. Higher oil prices damage the U.S. economy by transferring U.S. wealth to oil exporters, reducing real economic output, and creating temporary price and wage dislocations that lead to underemployment of economic resources. While the economic impact of the 1999–2000 oil price shock may have been smaller than the price shocks of the 1970s and 1980s, it was one of several factors causing a decline in U.S. economic growth in 2000 and 2001.

By reducing U.S. petroleum demand, greater fuel economy for passenger cars and light trucks ameliorates but does not by itself solve the problem of oil dependence. Because the United States accounts for 25 percent of world petroleum consumption (EIA, 2000c, table 11.9), changes in U.S. oil demand can significantly affect world oil prices. The size of the impact will depend on the price elasticity of net oil supply to the United States (Greene and Tishchishyna, 2000). A reasonable range of estimates of this elasticity is approximately 2.0 to 3.0, which means that a 1 percent decrease in U.S. demand would reduce world oil prices by 0.5 to 0.33 percent.

U.S. oil consumption in 2000 was 19.5 mmbd, so that the estimated 2.8-mmbd reduction due to fuel economy improvements represents a 13 percent reduction in U.S. oil demand, using the midpoint formula. Using the above elasticity assumptions, this should reduce world oil prices by 4 to 6 percent. The average price of oil in 2000 was just over $28/bbl, implying a savings of $1.00 to $1.80/bbl on every barrel purchased.

Accordingly, the reduction in expenditures realized by the United States due to lower prices for imported oil—which in turn came from improvements in passenger car and light-truck fuel economy—would be in the range of $3 billion to $6 billion for the year 2000 alone. This is in addition to the benefit of not having had to purchase those 2.8 mmbd. Assuming that these benefits increased linearly from zero in 1975, cumulative (not present value) oil-market benefits would amount to between $40 billion and $80 billion. This does not take into account benefits accruing from a reduced likelihood and severity of oil market disruptions. These estimates are subject to considerable uncertainty, however, because it is difficult to accurately predict OPEC responses to changes in oil demand.

The committee emphasizes again that these impacts on oil consumption and oil prices were the result of several factors affecting the fuel economy of the U.S. light-duty vehicle fleet, one of which was CAFE standards.

Regulatory Issues

In addition to the above issues, the committee was asked in the statement of work to address other aspects of how CAFE has functioned. These included the disparate impact on automotive manufacturers, the distinction between cars and light trucks, and the distinction between domestic and imported vehicle fleets.

Disparate Impacts of CAFE Standards

Some degree of differential or disparate impacts is inherent in a regulatory standard that sets the same performance measure for all manufacturers regardless of the type of vehicles they produce. Differences in the sizes and weights of domestically manufactured and imported vehicles in 1975 were described above. As Figure 2-2 illustrates, the domestic manufacturers (that is, Chrysler, GM, and Ford) had to improve the fuel economy of their vehicle fleets substantially, while foreign manufacturers (for example, Honda, Nissan, and Toyota) were already above the standards. Thus, some companies were affected to a greater extent than others. There is no doubt that the requirement to focus resources on the task of improving fuel economy called for greater investments and resources diverted from activities the domestic manufacturers would otherwise have preferred to pursue. Whether, in the end, this was harmful to U.S. manufacturers is less clear. Some argue that the fuel economy standards actually put U.S. manufacturers in a better competitive position when the oil price shocks hit in 1979 and 1980 than they would have been in had they been allowed to respond to falling gasoline prices between 1974 and 1978.

Passenger Cars and Light Trucks

The CAFE standards called for very different increases in passenger car and light-truck fuel economy. Passenger-car standards required a 75 percent increase from the new car fleet average of 15.8 mpg in 1975 to 27.5 mpg in 1985. Light-truck standards required only a 50 percent increase: from 13.7 mpg in 1975 to 20.7 mpg in 1987.

In part, the difference was intentional, reflecting the belief that light trucks function more as utility vehicles and face more demanding load-carrying and towing requirements. It was also due to the different mechanisms Congress established for setting the standards. Congress itself wrote the 27.5-mpg passenger-car target into law, while light-truck targets were left to the NHTSA to establish via rule-makings. The result of this process was that passenger cars were required to make a significantly greater percentage improvement in fuel economy.

The Foreign/Domestic Distinction

Automotive manufacturing is now a fully global industry. In 1980 the United Auto Workers (UAW) had 1,357,141 members, most of whom were employed in the automotive industry. However, by 2000 that number had dropped to 728,510 members, according to the annual report filed by the UAW with the Department of Labor. The loss of market share to foreign manufacturers, including some 35,000 assembly jobs in foreign-owned assembly plants in the United States, improvements in productivity in domestic plants, and a shift of parts production to Mexico as well as to nonunion foreign-owned parts plants in the United States resulted in the loss of unionized automotive jobs in the United States. Workers in this country have proven that they can compete successfully with workers overseas in all segments of the market, from the smallest cars to the largest trucks. The 1992 NRC report found that the provision of the CAFE law that created a distinction between domestic and foreign fleets led to distortions in the locations at which vehicles or parts are produced, with no apparent advantage (NRC, 1992, p. 171). NHTSA eliminated the domestic/import distinction for light trucks after model year 1995. The absence of negative effects of this action on employment in U.S. automobile manufacturing suggests that the same could be done for automobiles without fear of negative consequences.

Other Regulations Affecting CAFE

The gas guzzler tax, which first took effect in 1980, specifies a sliding tax scale for new passenger cars getting very low gas mileage. There is no comparable tax for light trucks. The level at which the tax takes effect increased from 14.5 mpg in 1980 to 22.5 mpg today, and the size of the tax has increased substantially. Today, the tax on a new passenger car achieving between 22 and 22.5 mpg is $1,000, increasing to $7,700 for a car with a fuel economy rating under 12.5 mpg. In 1975, 80 percent of new cars sold achieved less than 21 mpg and 10 percent achieved less than 12 mpg. In 2000, only 1 percent of all cars sold achieved less than 21.4 mpg (EPA, 2000). The tax, which applies only to new automobiles, has undoubtedly reinforced the disincentive to produce inefficient automobiles and probably played a role, as did the CAFE standards, in the downsizing of the passenger car fleet. The absence of a similar tax for light trucks has almost certainly exacerbated the disparities between the two vehicle types.

Emissions

Since the passage of the CAFE law in 1975, pollutant emissions standards for passenger cars and light trucks have been tightened. For example, hydrocarbon, carbon monoxide (CO), and nitrogen oxide (NO_x) federal standards were 1.5, 15, and 3.1 grams/mile, respectively, in 1975. Under Tier 1 standards, the analogous standards for nonmethane hydrocarbons, CO, and NO_x are 0.25, 3.4, and 0.4 grams/mile (Johnson, 1988; P.L.101-549). Moreover, the period for which new vehicles must be certified to perform effectively was doubled. The CAFE standards did not interfere with the implementation of emissions control standards. Indeed, several key fuel economy technologies are also essential for meeting today's emissions standards, and fuel economy improvements have been shown to help reduce emissions of hydrocarbons (Greene et al., 1994; Harrington, 1997). However, emissions standards have so far prevented key fuel economy technologies, such as the lean-burn gasoline engine or the diesel engine, from achieving significant market shares in U.S. light-duty vehicle markets.

Safety

Since 1975, many new passenger car and light-truck safety regulations have been implemented. It was estimated that these regulations added several hundred pounds to the average vehicle (for example, air bags and improved impact protection). However, the actual number may now be less (there have not been any follow-up studies to determine if improved designs and technological progress have reduced the weight of those components). Nonetheless, the CAFE regulations, have not impeded the implementation of safety regulations and safety regulations have not prevented manufacturers from achieving their CAFE requirements.

IMPACTS ON THE AUTOMOBILE INDUSTRY

Regulations such as the CAFE standards are intended to direct some of industry's efforts toward satisfying social goals that transcend individual car buyers' interests. Inevitably, they divert effort from the companies' own goals. This section reviews trends in revenues, profits, employment, R&D spending, and capital investment for the domestic automobile industry from 1972 to 1997. Examination of the data shows little evidence of a dramatic impact of fuel economy regulations. General economic conditions, and especially the globalization of the automobile industry, seem to have been far more important than fuel economy regulations in determining the profitability and employment shares of the domestic automakers and their competitors.

The 1992 NRC report on automobile fuel economy concluded, "Employment in the U.S. automotive industry has declined significantly and the trend is likely to continue during the 1990s. The world automotive industry, particularly the domestic industry, suffers from over-capacity, and further plant closings and reductions in employment are inevitable" (NRC, 1992). Fortunately, this gloomy prediction turned out to be largely incorrect, as total employment in automobile manufacturing in the United States reached its highest level ever (more than 1 million) in 1999 (Figure 2-10), thanks largely to foreign companies' decisions to move manufacturing to the United States to take advantage of the most profitable market in the world. In 1990 there were eight foreign-owned plants in the United States producing 1.49 million vehicles annually. By 2000, foreign companies assembled 2.73 million vehicles in 11 U.S. plants; Honda and Nissan will each open another new assembly plant in the next 2 years.

Organized labor has lost nearly half of its representation in the automobile industry since 1980. In that year, the United Auto Workers union had 1.4 million members, most of them employed in the auto industry, but by year-end 2000 it reported just over 670,000 members (UAW, 2000). The roots of this shift include the domestic manufacturers' loss of market share to foreign manufacturers, improved productivity in their own plants, and shifts of parts production overseas. The job losses have been offset by about 35,000 jobs in foreign-owned, nonunion assembly plants in the United States; growth in white collar employment in foreign companies as they expanded distribution; and the establishment of foreign-owned parts and component operations.

Like profitability, two measures of productivity show no obvious impact of fuel economy improvements. The number of light-duty vehicles produced per worker (Figure 2-10) has fluctuated with the business cycle (falling during recessions) and since the mid-1990s appears to have trended slightly upward despite increased production of light trucks and more complex cars. The sales value of cars produced per worker (also shown in Figure 2-10) increased substantially during the 1972 to 2000 period, particularly after 1980.

Even before the CAFE standards were established, the automotive market was becoming a global one. In the 1960s imported vehicles made significant inroads in the United States. With their small cars and their reputation for superior quality, Japanese producers probably found the CAFE standards only one source of competitive advantage in U.S. markets among others during the 1970s and 1980s. The size of this advantage is difficult to determine, however (NRC, 1992).

The industry's ability to fund R&D and capital investment is a function of its financial strength. The annual net

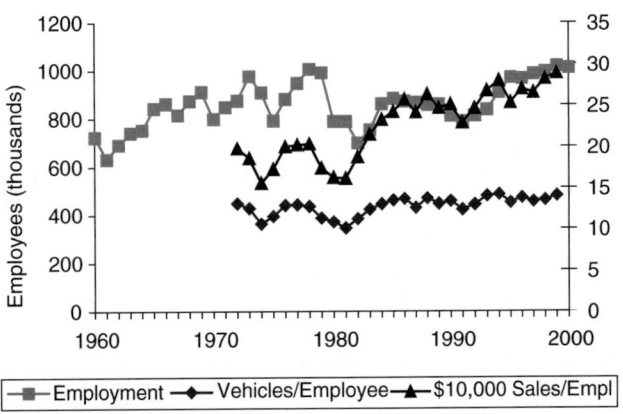

FIGURE 2-10 Employment and productivity in the U.S. automotive industry. SOURCE: *Wards Automotive Report*.

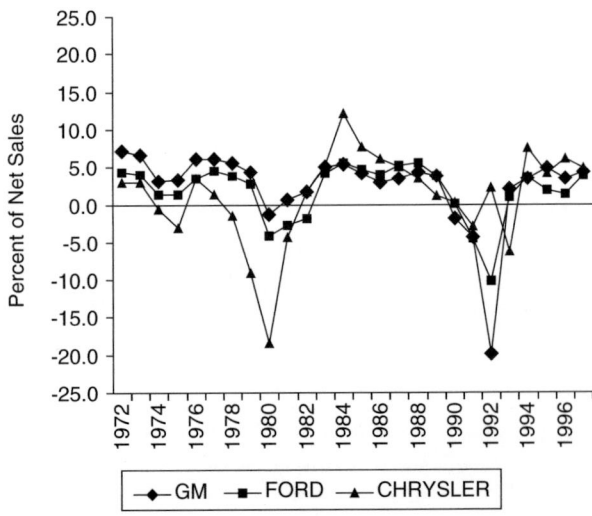

FIGURE 2-11 Net profit rates of domestic manufacturers, 1972–1997. SOURCE: *Wards Automotive Report*.

income of GM, Ford, and Chrysler is correlated closely with the business cycle and the competitiveness of each company's products (Figure 2-11). The industry experienced severe losses in 1980 and again in 1992 in response to the drop in vehicle demand, a competitive pricing environment, and loss of market share to foreign producers. After that, the industry enjoyed a powerful rebound in earnings. Between 1994 and 1999, the cumulative net income of GM, Ford, and Chrysler amounted to an all-time record of $93 billion.

The most important cause of this rebound was exploding demand for light trucks, a market sector dominated by the Big Three:

- In 1984 minivans were introduced and by 1990 were selling nearly a million units annually.
- Then came four-door SUVs and pickup trucks with passenger-friendly features such as extra rows of seats. SUV sales increased from fewer than 1 million units in 1990 to 3 million in 2000; large SUVs were the fastest-growing segment and by 2000 accounted for nearly one-third of all SUVs sold. Sales of large pickup trucks nearly doubled in the 1990s.
- Crossover vehicles, which have trucklike bodies on car platforms, offer consumers an alternative to a traditional light truck. These vehicles (for example, the Toyota RAV-4 and the Honda CRV), first introduced by Japanese companies several years ago to serve demand for recreational vehicles, also found markets in the United States. U.S. auto companies are now launching models in this category.

Light trucks today account for about 50 percent of GM sales, 60 percent of Ford sales, and 73 percent of DaimlerChrysler sales and even greater shares of the profits of all three companies. In the mid- to late 1990s, the average profit on a light truck was three to four times as great as that on a passenger sedan.

Since the second half of 2000, however, GM and Ford have recorded sharply lower profits, and the Chrysler division of DaimlerChrysler suffered significant losses. A slowing economy, which necessitated production cuts as well as purchase incentives (rebates and discounted loan rates, for example) to defend market share, underlies the downturn in industry profitability.

With at least 750,000 units of additional capacity of light-truck production coming onstream over the next 3 years, however, margins on these vehicles could remain under pressure for the foreseeable future. To recoup their investments in truck capacity, manufacturers will continue to use incentives to drive sales, even at the cost of lower unit profits. (Better incentives have made these vehicles more affordable, which probably explains some of their continuing popularity in the face of higher fuel prices.)

Two important indicators of the costs of the CAFE standards to industry, regardless of its profitability, are the investments required in changing vehicle technology and design: Investments for retooling and R&D must be recovered over time in the profits from vehicle sales. Too steep an imposition of the standards would be reflected in unusually high rates of both investments. There does appear to have been a sudden increase in retooling investments by all three manufacturers in 1980, but they returned to normal levels within 5 years (see Figure 2-12). These investments may have been prompted as much by changes in U.S. manufacturers' market strategies as by the impending CAFE standards (their strategies may, for example, have reflected expectations that fuel prices would continue rising steeply and

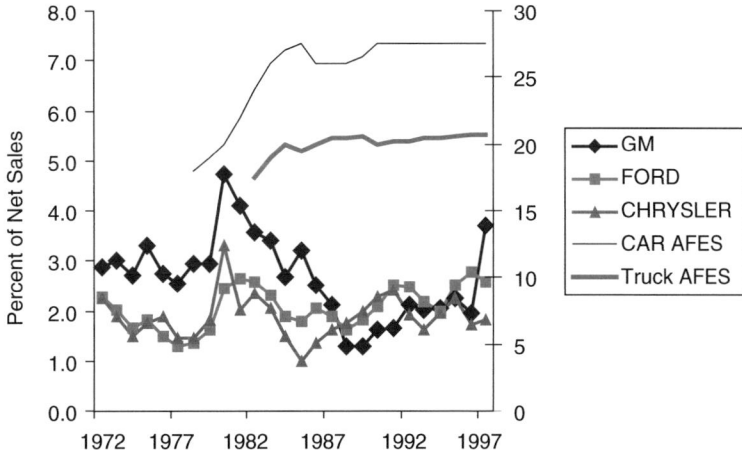

FIGURE 2-12 Investments in retooling by domestic automobile manufacturers, 1972–1997, with automotive fuel economy standards (AFES) for passenger cars and trucks. SOURCE: DOT Docket 98-4405-4. Advanced Air Bag Systems Cost, Weight, and Lead Time Analysis Summary Report, Appendix A (Contract No. DTNH22-96-0-12003, Task Orders-001, 002, and 003).

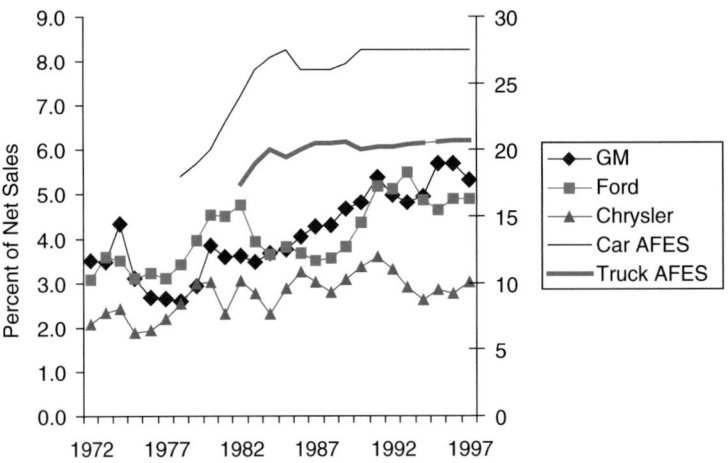

FIGURE 2-13 R&D investments by domestic automobile manufacturers, 1972–1997, with automotive fuel economy standards (AFES) for passenger cars and trucks. SOURCE: DOT Docket 98-4405-4. Advanced Air Bag Systems Cost, Weight, and Lead Time Analysis Summary Report, Appendix A (Contract No. DTNH22-96-0-12003, Task Orders-001, 002, and 003).

the need in general to counter the Japanese competition's reputation for quality).

Investments in R&D as a percent of net sales were relatively low in the years leading up to 1978, the first year in which manufacturers were required to meet the CAFE standards. Since then, they have generally increased, regardless of whether the CAFE requirements were increasing or constant (Figure 2-13). This pattern suggests that the R&D demands created by the standards did not unduly burden the domestic manufacturers.

IMPACT ON SAFETY

In estimating the effect of CAFE on safety, the committee relied heavily on the 1992 NRC report *Automotive Fuel Economy: How Far Should We Go?* (NRC, 1992). That report began its discussion of the safety issues by noting, "Of all concerns related to requirements for increasing the fuel economy of vehicles, safety has created the most strident public debate" (NRC, 1992, p. 47).

Principally, this debate has centered on the role of vehicle mass and size in improving fuel economy. For a given power train, transportation fuel requirements depend in part on how much mass is moved over what distance, at what speed, and against what resistance. The mass of the vehicle is critical because it determines the amount of force (that is, power and fuel) necessary to accelerate the vehicle to a given speed or propel it up a hill. Size is important because it influences mass (larger vehicles usually weigh more) and, secondarily, because it can influence the aerodynamics of the vehicle and, therefore, the amount of power necessary to keep it moving at a given speed.

As discussed above, fuel economy improved dramatically for cars during the late 1970s and early 1980s, without much change since 1988 (see Figure 2-2 and Figure 2-6). That increase in fuel economy was accompanied by a decline in average car weight (see Figure 2-5) and in average wheelbase length (a common measure of car size). Thus, a significant part of the increased fuel economy of the fleet in 1988 compared with 1975 is attributable to the downsizing of the vehicle fleet. Since 1988, new cars have increased in weight (see Figure 2-5) and the fuel economy has suffered accordingly (see Figure 2-4), although increasing mass is not the only reason for this decline in fuel economy.

The potential problem for motor vehicle safety is that vehicle mass and size vary inversely not only with fuel economy, but also with risk of crash injuries. When a heavy vehicle strikes an object, it is more likely to move or deform the object than is a light vehicle. Therefore the heavier vehicle's occupants decelerate less rapidly and are less likely to be injured. Decreasing mass means that the downsized vehicle's occupants experience higher forces in collisions with other vehicles. Vehicle size also is important. Larger crush zones outside the occupant compartment increase the distance over which the vehicle and its restrained occupants are decelerated. Larger interiors mean more space for restraint systems to effectively prevent hard contact between the occupants' bodies and the structures of the vehicle. There is also an empirical relationship, historically, between vehicle mass/size and rollover injury likelihood. These basic relationships between vehicle mass, size, and safety are discussed in greater detail in Chapter 4.

What Has Been the Effect of Changes in Vehicle Mass and Size on Motor Vehicle Travel Safety?

Given these concerns about vehicle size, mass, and safety, it is imperative to ask about the safety effect of the vehicle downsizing and downweighting that occurred in association with the improvement in fuel economy during the 1970s and 1980s. There are basically two approaches to this question. Some analysts have concluded that the safety effect of fleet downsizing and downweighting has been negligible because the injury and fatality experience per vehicle mile of travel has declined steadily during these changes in the fleet. The General Accounting Office (GAO) championed this view in a 1991 report, arguing that vehicle downweighting and downsizing to that time had resulted in no safety consequences, as engineers had been able to offset any potential risks (Chelimsky, 1991). According to this argument, the fact that vehicle downsizing and downweighting have not led to a large increase in real-world crash injuries indicates that there need not be a safety penalty associated with downsizing, despite any theoretical or empirical relationships among the size, weight, and safety of vehicles at a given time.

As the 1992 NRC report indicated, however, that view has been challenged (NRC, 1992, pp. 54–55). The reduced risk of motor vehicle travel during the past decade is part of a long-term historical trend, going back to at least 1950 (Figure 2-14). Most important, the improving safety picture is the result of various interacting—and, sometimes, conflicting—trends. On the one hand, improved vehicle designs, reduced incidence of alcohol-impaired driving, increased rates of safety belt use, and improved road designs are reducing crash injury risk; on the other, higher speed limits, increased horsepower, and increasing licensure of teenagers and other risky drivers, among other factors, are increasing crash injury risk. In short, the historical trend in motor vehicle injury and fatality rates is too broad a measure, affected by too many variables, to indicate whether vehicle downsizing and downweighting have increased or decreased motor vehicle travel safety.

Recognizing this general historical trend, the appropriate question is not whether crash injury risk has continued to decline in the face of vehicle downsizing and downweighting, but rather whether motor vehicle travel in the downsized fleet is less safe than it would have been otherwise. This approach to the question treats the safety charac-

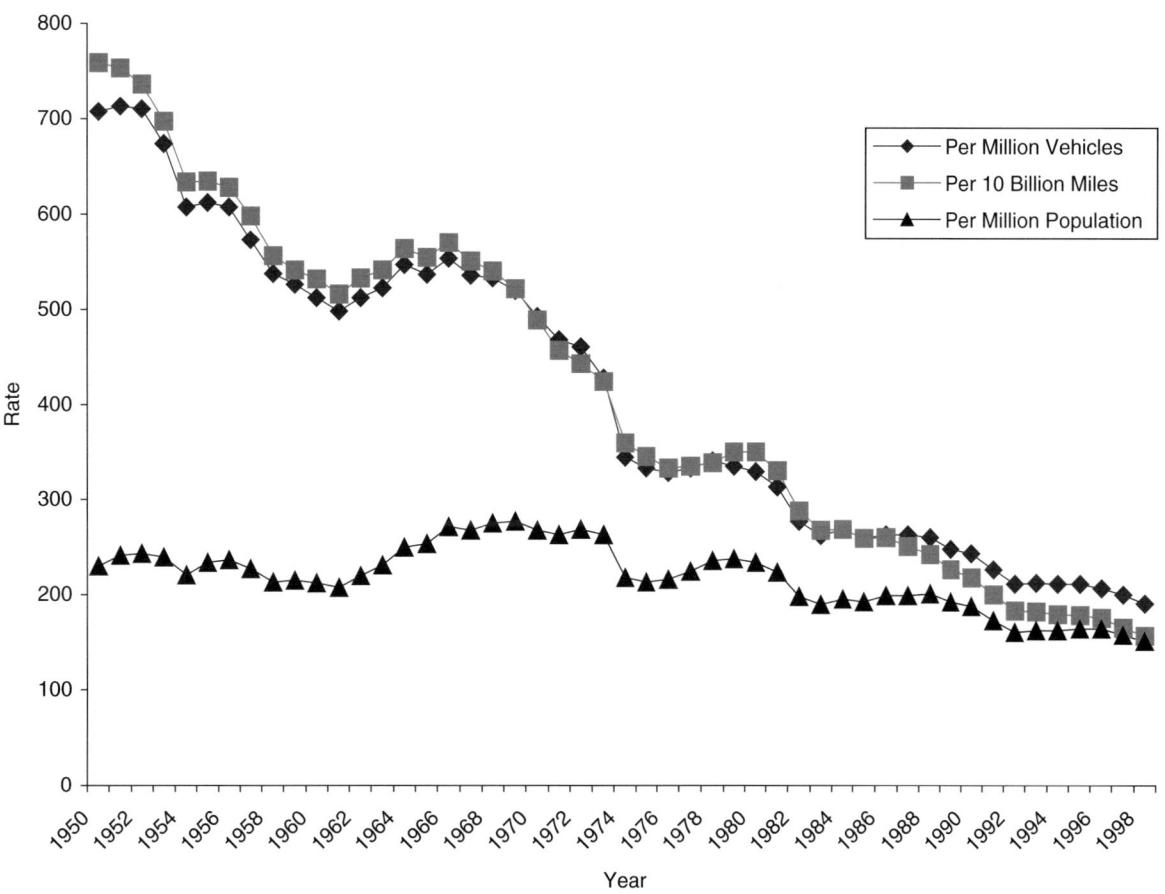

FIGURE 2-14 Motor vehicle crash death rates, 1950–1998. SOURCE: National Safety Council (1999).

teristics of the motor vehicle fleet at any particular time as a given. That is, the level of safety knowledge and technology in use at the time is independent of the size and weight of the vehicle fleet. Accordingly, the question for evaluating the safety effects of constraints on vehicle size and weight asks how much injury risk would change if consumers were to purchase larger, heavier vehicles of the generation currently available to them.

The 1992 NRC report noted significant evidence that the improvement in motor vehicle travel safety to that time could have been even greater had vehicles not been downweighted and downsized. For example, the report cited NHTSA research (Kahane, 1990; Kahane and Klein, 1991) indicating that "the reductions that have occurred in passenger-vehicle size from model year 1970 to 1982 are associated with approximately 2,000 additional occupant fatalities annually" (NRC, 1992, p. 53). In another study cited by the 1992 report, Crandall and Graham (1988) estimated that fatality rates in 1985 car models were 14 to 27 percent higher because of the 500 lb of weight reduction attributed by those authors to CAFE requirements. These estimates revealed forgone reductions in fatalities occasioned by the downweighting and/or downsizing of the fleet. These safety costs had been hidden from public view by the generally improving safety of the motor vehicle environment.

It should be noted that the terms *downsizing* and *downweighting* are used interchangeably here because of the very high correlation between these physical attributes of motor vehicles. Although the effects of size and mass appear quite separate in the theoretical discussion above, in reality most heavy cars are large and most large cars are heavy. As a result of this correlation, the 1992 NRC report was unable to separate the different effects of vehicle mass and size in accounting for the changes in safety. The report questioned to what extent the increased fatalities due to downweighting could have been prevented had vehicles retained their initial size.

Nevertheless, the report concluded that "the historical changes in the fleet—downsizing and/or downweighting—have been accompanied by increased risk of occupant injury" (NRC, 1992, p. 55). The current committee concurs with that conclusion.

Societal Versus Individual Safety

The 1992 NRC report also questioned the relationship between risk to the individual occupant of downsized vehicles and risk to society as a whole. Specifically, the report questioned whether estimates of the effects of downsizing adequately assessed "the net effects of the safety gains to the occupants of the heavier car and safety losses that the increased weight imposes on the occupants of the struck car, as well as other road users (e.g., pedestrians, pedalcyclists, and motorcyclists)" (NRC, 1992, p. 57). In other words, larger mass means greater protection for the occupants of the vehicle with greater mass but greater risk for other road users in crashes. Some of the increased risk for individuals shifting to smaller, lighter cars would be offset by decreased risk for individuals already in such cars. However, the report noted that there was insufficient information about the effects on all road users of changes in fleet size and weight distributions. It also noted that increasing sales of light trucks, which tend to be larger, heavier, and less fuel efficient than cars, was a factor increasing the problem of crash incompatibility. NHTSA was urged to conduct a study to develop more complete information on the overall safety impact of increased fuel economy and to incorporate more information about the safety impact of light-truck sales.

In April 1997, NHTSA issued a report summarizing research undertaken by it in response to that issue as well as to

TABLE 2-1 Change in Death or Injury Rates for 100-lb Weight Reduction in Average Car or Average Light Truck (percent)

Crash Type	Fatality Analysis		Injury Analysis	
	Cars	Light Trucks	Cars	Light Trucks
Hit object	+1.12	+1.44	+0.7	+1.9
Principal rollover	+4.58	+0.81*	NE	NE
Hit passenger car	−0.62*	−1.39	+2.0	−2.6
Hit light truck	+2.63	−0.54*	+0.9	—
Hit big truck	+1.40	+2.63	—	
Hit ped/bike/motorcycle	−0.46	−2.03	NE	NE
Overall	+1.13	−0.26	+1.6	−1.3

NOTE: For the injury analysis, NE means this effect was not estimated in the analysis. A dash indicates the estimated effect was statistically insignificant. For the fatality analysis, the starred entries were not statistically significant.
SOURCE: NHTSA (1997).

other informational concerns expressed in the 1992 NRC report. In the new NHTSA research, the effect on fatalities and injuries of an average 100-lb reduction in the weight of cars (or in the weight of light trucks) was estimated. Following the recommendation of the 1992 NRC report, the fatality analysis included fatalities occurring to nearly all road users in crashes of cars and light trucks; excluded were only those fatalities occurring in crashes involving more than two vehicles and other rare events. The injury analysis was more limited, including only those injuries occurring to occupants of the cars and light trucks. Table 2-1 summarizes the NHTSA results.

NHTSA's fatality analyses are still the most complete available in that they accounted for all crash types in which vehicles might be involved, for all involved road users, and for changes in crash likelihood as well as crashworthiness. The analyses also included statistical controls for driver age, driver gender, and urban-rural location, as well as other potentially confounding factors. The committee's discussions focused on the fatality analyses, although the injury analyses yielded similar results to the extent that their limitations permitted comparison.

The NHTSA fatality analyses indicate that a reduction in mass of the passenger car fleet by 100 lb with no change in the light-truck fleet would be expected to increase fatalities in the crashes of cars by 1.13 percent. That increase in risk would have resulted in about 300 (standard error of 44) additional fatalities in 1993. A comparable reduction in mass of the light-truck fleet, with no change in cars, would result in a net reduction in fatalities of 0.26 percent (or 40 lives saved, with a standard error of 30) in 1993. NHTSA attributed this difference in effect to the fact that the light-truck fleet is on average 900 lb heavier than the passenger car fleet. As a result, the increased risk to light-truck occupants in some crashes as a result of downweighting is offset by the decreased risk to the occupants of other vehicles involved in collisions with them, most of which are much lighter. The results of the separate hypothetical analyses for cars and light trucks are roughly additive, so that a uniform reduction in mass of 100 lb for both cars and light trucks in 1993 would be estimated to have resulted in about 250 additional fatalities. Conversely, a uniform increase in mass of 100 lb for both cars and light trucks would be estimated to result in about 250 lives saved.

The April 1997 NHTSA analyses allow the committee to reestimate the approximate effect of downsizing the fleet between the mid-1970s and 1993. In 1976, cars were about 700 lb heavier than in 1993; light trucks were about 300 lb heavier, on average.[5] An increase in mass for cars and light-duty trucks on the road in 1993, returning them to the average weight in 1976, would be estimated to have saved about 2,100 lives in car crashes and cost about 100 fatalities in light-truck crashes. The net effect is an estimated 2,000 fewer fatalities in 1993, if cars and light trucks weighed the same as in 1976. The 95 percent confidence interval for this estimate suggests that there was only a small chance that the safety cost was smaller than 1,300 lives or greater than 2,600 lives. This figure is comparable to the earlier NHTSA estimates of the effect of downsizing since the early 1970s.

In short, even after considering effects on all road users and after adjusting the results for a number of factors known to correlate with both fatal crash risk and vehicle usage patterns, the downsizing and downweighting of the vehicle fleet that occurred during the 1970s and early 1980s still appear to have imposed a substantial safety penalty in terms of lost lives and additional injuries. The typical statistical relationship between injuries and fatalities in the NHTSA's accident data suggests that these changes in the fleet were responsible for an additional 13,000 to 26,000 incapacitating injuries and 97,000 to 195,000 total injuries in 1993.

It must be noted that the application of the 1997 NHTSA analyses to the questions before this committee is not without controversy. In 1996, after reviewing a draft of the NHTSA analyses, a committee of the National Research Council's Transportation Research Board (NRC-TRB) expressed concerns about the methods used in these analyses and concluded, in part, "the Committee finds itself unable to endorse the quantitative conclusions in the reports about projected highway fatalities and injuries because of large uncertainties associated with the results. . . ." These reservations were principally concerned with the question of whether the NHTSA analyses had adequately controlled for confounding factors such as driver age, sex, and aggressiveness. Two members of the current committee are convinced that the concerns raised by the NRC-TRB committee are still valid and question some of the conclusions of the NHTSA analyses. Their reservations are detailed in a dissent that forms Appendix A of this report.

The majority of the committee shares these concerns to an extent, and the committee is unanimous in its agreement that further study of the relationship between size, weight, and safety is warranted. However, the committee does not agree that these concerns should prevent the use of NHTSA's careful analyses to provide some understanding of the likely effects of future improvements in fuel economy, if those improvements involve vehicle downsizing. The committee notes that many of the points raised in the dissent (for example, the dependence of the NHTSA results on specific estimates of age, sex, aggressive driving, and urban vs. rural location) have been explicitly addressed in Kahane's response to the NRC-TRB review and were reflected in the final 1997 report. The estimated relationship between mass and safety were remarkably robust in response to changes in the estimated effects of these parameters. The committee also

---

[5]The average weights of cars and light trucks registered for use on the road in 1976 were, respectively, 3,522 lb and 3,770 lb; in 1993, 2,816 lb and 3,461 lb. The Insurance Institute for Highway Safety computed these weights, using R.L. Polk files for vehicle registration in those years and institute files on vehicle weights.

notes that the most recent NHTSA analyses (1997) yield results that are consistent with the agency's own prior estimates of the effect of vehicle downsizing (Kahane, 1990; Kahane and Klein, 1991) and with other studies of the likely safety effects of weight and size changes in the vehicle fleet (Lund and Chapline, 1999). This consistency over time and methodology provides further evidence of the robustness of the adverse safety effects of vehicle size and weight reduction.

Thus, the majority of this committee believes that the evidence is clear that past downweighting and downsizing of the light-duty vehicle fleet, while resulting in significant fuel savings, has also resulted in a safety penalty. In 1993, it would appear that the safety penalty included between 1,300 and 2,600 motor vehicle crash deaths that would not have occurred had vehicles been as large and heavy as in 1976.

Changes in the Fleet Since 1993

As noted earlier, vehicle weights have climbed slightly in recent years, with some regressive effects on vehicle fuel economy. The committee sought to estimate the effect of these later changes on motor vehicle safety, as well. However, there is some uncertainty in applying NHTSA's estimates directly to fatal crash experience in other years. First, it is possible that the safety effects of size and weight will change as vehicle designs change; for example, it is possible that substitution of lighter-weight structural materials could allow vehicles to reduce weight while maintaining protective size to a greater extent than in the past. Second, the effects of vehicle size and weight vary for different crash types, as noted in Table 2-1, and the frequency distribution of these crash types can vary from year to year for reasons other than vehicle size and weight. Therefore, one needs to know this distribution before one can apply NHTSA's estimates.

Historical Relationships Between Size or Weight and Occupant Protection

Whether the safety effects of size and weight change as vehicles are redesigned can ultimately be determined definitively only by replication of NHTSA analyses (Kahane, 1997). However, a review of the historical relationship between size, weight, and occupant protection indicates that the risk reduction associated with larger size and weight has been reasonably stable over the past 20 years. For example, Table 2-2 shows occupant death rates in different light-duty vehicle classes for 1979, 1989, and 1999 (the last year for which federal data on fatalities are available). The data show that fatality rates per registered vehicle improved among all vehicle type and size/weight classes between 1979 and 1999, but the ratio of fatality risk in the smallest vehicles of a given type compared with the largest did not change much. The single exception has been among small utility vehicles, where there was dramatic improvement in the rollover fatality risk between 1979 and 1989. In short, although it is possible that the weight, size, and safety relationships in future vehicle fleets *could* be different from those in the 1993 fleet

TABLE 2-2 Occupant Deaths per Million Registered Vehicles 1 to 3 Years Old

Vehicle Type	Vehicle Size	Year		
		1979	1989	1999
Car	Mini	379	269	249
	Small	313	207	161
	Midsize	213	157	127
	Large	191	151	112
	Very large	160	138	133
	All	244	200	138
Pickup	<3,000 lb	384	306	223
	3,000–3,999 lb	314	231	180
	4,000–4,999 lb	256	153	139
	5,000+ lb	—	94	115
	All	350	258	162
SUV	<3,000 lb	1,064	192	195
	3,000–3,999 lb	261	193	152
	4,000–4,999 lb	204	111	128
	5,000+ lb	—	149	92
	All	425	174	140
All passenger vehicles		265	208	143

NOTE: Cars are categorized by wheelbase length rather than weight. SOURCE: Insurance Institute for Highway Safety, using crash death data from the Fatality Analysis Reporting System (NHTSA) and vehicle registration data from R.L. Polk Company for the relevant years.

TABLE 2-3 Distribution of Motor Vehicle Crash Fatalities in 1993 and 1999 by Vehicle and Crash Type

Vehicle	Crash Type	Year 1993	Year 1999	Percent Change
Car	Principal rollover	1,754	1,663	−5
	Object	7,456	7,003	−6
	Ped/bike/motorcycle	4,206	3,245	−23
	Big truck	2,648	2,496	−6
	Another car	5,025	4,047	−19
	Light truck	5,751	6,881	+20
Light truck	Principal rollover	1,860	2,605	+40
	Object	3,263	3,974	+22
	Ped/bike/motorcycle	2,217	2,432	+10
	Big truck	1,111	1,506	+36
	Car	5,751	6,881	+20
	Another light truck	1,110	1,781	+60
Total/average		36,401	37,633	+3

SOURCE: The programs for counting the relevant crash fatality groups were obtained from Kahane and applied to the 1999 Fatality Analysis Reporting System (FARS) file by the Insurance Institute for Highway Safety.

studied by Kahane (1997), there appears to be no empirical reason to expect those relationships *will* be different. Thus, the majority of the committee believes that it is reasonable to use the quantitative relationships developed by NHTSA (Kahane, 1997) and shown in Table 2-1 to estimate the safety effects of vehicle size and weight changes in other years.

Distribution of Crash Types in the Future

While there appears to be some justification for expecting relationships among weight, size, and safety to remain much the same in the future, the committee observed that, between 1993 and 1999, the last year for which complete data on fatal crashes are available, there were several shifts in fatal crash experience, the most notable being an increase in the number of light-duty truck involvements (consistent with their increasing sales) and a decrease in crashes fatal to nonoccupants (pedestrians and cyclists; see Table 2-3). The result of these changes in crash distribution is that the estimated effect on all crash fatalities of a 100-lb gain in average car weight increased, from −1.13 percent in 1993 (Kahane, 1997) to −1.26 percent in 1999; the estimated effect on crash fatalities of a 100-lb gain in average light-truck weight decreased from +0.26 percent in 1993 to +0.19 percent in 1999.

Between 1993 and 1999, the average weight of new passenger cars increased about 100 lb, and that of new light trucks increased about 300 lb.[6] The results in the preceding paragraph suggest that the fatality risk from all car crashes has declined as a result of this weight gain by 1.26 percent (or about 320 fewer fatalities), while the net fatality risk from light-truck crashes has increased by 0.57 percent (or about 110 additional fatalities). The net result is an estimated 210 fewer deaths in motor vehicle crashes of cars and light trucks (or between 10 and 400 with 95 percent confidence). Thus, the indications are that recent increases in vehicle weight, though detrimental to fuel economy, have saved lives in return.

The preceding discussion has acknowledged some uncertainty associated with the safety analyses that were reviewed in the preparation of this chapter. Based on the existing literature, there is no way to apportion precisely the safety impacts, positive or negative, of weight reduction, size reduction, vehicle redesign, and so on that accompanied the improvements in fuel economy that have occurred since the mid-1970s. But it is clear that there were more injuries and fatalities than otherwise would have occurred had the fleet in recent years been as large and heavy as the fleet of the mid-1970s. To the extent that the size and weight of the fleet have been constrained by CAFE requirements, the current committee concludes that those requirements have caused more injuries and fatalities on the road than would otherwise have occurred. Recent increases in vehicle weight, while resulting in some loss of fuel economy, have probably resulted in fewer motor vehicle crash deaths and injuries.

REFERENCES

Chelimsky, E. 1991. Automobile Weight and Safety. Statement before the Subcommittee on Consumers. Committee on Commerce, Science, and

---

[6]In 1999, the average weight of cars registered was 2,916 lb; for trucks, 3,739 lb. See footnote 5 also.

Transportation. U.S. Senate. GAO/T-PEMD-91-2. Washington, D.C.: U.S. General Accounting Office.

Crandall, R.W., and J.D. Graham. 1988. The Effect of Fuel Economy Standards on Automobile Safety. The Brookings Institution and Harvard School of Public Health. NEIPRAC Working Paper Series, No. 9. Cambridge, Mass.: New England Injury Prevention Center.

Crandall, R.W., H.K. Gruenspecht, T.E. Keeler, and L.B. Lave. 1986. Regulating the Automobile. Washington, D.C.: The Brookings Institution.

Darby, M.R., 1982. "The Price of Oil and World Inflation and Recession." *American Economic Review* 72:738–751.

Davis, S.C. 2000. Transportation Energy Data Book: Edition 20. ORNL-6959. Oak Ridge, Tenn.: Oak Ridge National Laboratory.

DOE (Department of Energy). 1991. National Energy Strategy. DOE/S-0082P, February. Washington, D.C.: Department of Energy.

Eastwood, R.K. 1992. "Macroeconomic Impacts of Oil Price Shocks," *Oxford Economic Papers* 44: 403–425.

EIA (Energy Information Administration). 1998. What Does the Kyoto Protocol Mean to U.S. Energy Markets and the U.S. Economy? SR/OIAF/98-03. Washington, D.C. Available online at <http://www.eia.doe.gov/oiaf/kyoto/kyotobtxt.html>.

EIA. 2000a. Annual Energy Outlook 2001. DOE/EIA-0383(2001). Washington, D.C.: EIA.

EIA. 2000b. Emissions of Greenhouse Gases in the United States 1999. DOE/EIA-0573(99). Washington, D.C.: EIA.

EIA. 2000c. Oil Price Impacts on the U.S. Economy, presentation slides and notes, February 21, 2000. Available online at <ftp://ftp.eia.doe.gov/pub/pdf/feature/Econ1/sld001.htm>.

EPA (Environmental Protection Agency). 2000. Light-duty Automotive Fuel Economy Trends 1975 Through 2000. Office of Air and Radiation. EPA420-00-008. Ann Arbor, Mich.: EPA.

EPA. 2001. Inventory of U.S. Greenhouse Gas Emissions and Sinks: 1990–1999. Draft, January 8, 2001. Washington, D.C.: EPA. Available online at <http://www.epa.gov/globalwarming/publications/emissions/us2001/index.html>.

FHWA (Federal Highway Administration). 2000. Highway Statistics. Washington, D.C.: FWHA, Department of Transportation.

Greene, D.L. 1990. "CAFE or Price? An Analysis of the Effects of Federal Fuel Economy Regulations and Gasoline Price on New Car MPG, 1978–1989." *The Energy Journal* 11(3): 37–57.

Greene, D.L. 1998. "Why CAFE Worked." *Energy Policy* 26(8): 595–613.

Greene, D.L., and N.I. Tishchishyna. 2000. Costs of Oil Dependence: A 2000 Update. ORNL/TM-2000/152 (May). Oak Ridge, Tenn.: Oak Ridge National Laboratory.

Greene, D.L., M.A. Delucchi, and M. Wang. 1994. "Motor Vehicle Fuel Economy: The Forgotten Hydrocarbon Control Strategy?" *Transportation Research-A* 28A(3): 223–244.

Greene, D.L., D.W. Jones, and P.N. Leiby. 1998. "The Outlook for U.S. Oil Dependence. *Energy Policy* 26(1): 55–69.

Greene, D.L., J. Kahn, and R. Gibson. 1999. "Fuel Economy Rebound Effect for U.S. Household Vehicles." *The Energy Journal* 20(3): 1–31.

Hamilton, J.D. 1983. "Oil and the Macroeconomy Since World War II." *Journal of Political Economy* 91: 228–248.

Hamilton, J.D. 1996. Analysis of the Transmission of Oil Price Shocks Through the Economy. Presented at the Symposium on International Energy Security: Economic Vulnerability to Oil Price Shocks, Washington, D.C., October 3–4.

Harrington, W. 1997. "Fuel Economy and Motor Vehicle Emissions." *Journal of Environmental Economics and Management* 33: 240–252.

Harrison, I.M. 1996. RTECS Research Fuel Purchase Log Study. Memorandum. U.S. Department of Energy, Energy Information Administration (May 2), Washington, D.C.: EIA.

Haughton, J., and S. Sarker. 1996. "Gasoline Tax As a Corrective Tax: Estimates for the United States, 1970–1991." *The Energy Journal* 17(2): 103–126.

Hellman, K.H., and J.D. Murrell. 1984. Development of Adjustment Factors for the EPA City and Highway MPG Values. SAE Technical Paper Series 840496. Warrendale, Pa.: Society of Automotive Engineers.

Hickman, B.G. 1987. "Macroeconomic Impacts of Energy Shocks and Policy Responses: A Structural Comparison of Fourteen Models." *Macroeconomic Impacts of Energy Shocks*. B. Hickman, H. Huntington, and J. Sweeney, eds. Elsevier Science.

Huntington, H.G. 1996. Estimating Macroeconomic Impacts of Oil Price Shocks with Large-Scale Econometrics Models: EMF-7 and Recent Trends. Presented at the Symposium on International Energy Security: Economic Vulnerability to Oil Price Shocks, Washington, D.C., October 3–4.

IPCC (Intergovernmental Panel on Climate Change). 2001. Climate Change 2001: The Scientific Basis. Cambridge, U.K.: Cambridge University Press.

Johnson, J.H. 1988. "Automotive Emissions." In Air Pollution, the Automobile, and Public Health. Washington, D.C.: National Academy Press.

Jones, C.T. 1993. "Another Look at U.S. Passenger Vehicle Use and the 'Rebound' Effect from Improved Fuel Efficiency." *The Energy Journal* 14(4): 99–110.

Kahane, C.J. 1990. Effect of Car Size on Frequency and Severity of Rollover Crashes. Washington, D.C.: National Highway Traffic Safety Administration.

Kahane, C.J. 1997. Relationships Between Vehicle Size and Fatality Risk in Model Year 1985–93 Passenger Cars and Light Trucks. NHTSA Technical Report, DOT HS 808 570. Springfield, Va.: National Technical Information Services.

Kahane, C.J., and T. Klein. 1991. Effect of Car Size on Fatality and Injury Risk. Washington, D.C.: National Highway Traffic Safety Administration.

Leone, R.A., and T.W. Parkinson. 1990. Conserving Energy: Is There a Better Way? A Study of Corporate Average Fuel Economy Regulation. Washington, D.C.: Association of International Automobile Manufacturers.

Lund, A.K., and J.F. Chapline. 1999. Potential Strategies for Improving Crash Compatibility in the U.S. Vehicle Fleet. SAE 1999-01-0066. Vehicle Aggressivity and Compatibility in Automotive Crashes (SP-1442), 33-42. Warrendale, Pa.: Society of Automotive Engineers.

Mintz, M., A.D. Vyas, and L.A. Conley. 1993. Differences Between EPA-Test and In-Use Fuel Economy: Are the Correction Factors Correct? Transportation Research Record 1416, Transportation Research Board, National Research Council. Washington, D.C.: National Academy Press.

Mork, K.A., O. Olsen, and H.T. Mysen. 1994. "Macroeconomic Responses to Oil Price Increases and Decreases in Seven OECD Countries." *The Energy Journal* 15(4): 19–35.

National Safety Council. 1999. Injury Facts. Chicago, Ill.

NHTSA (National Highway Traffic Safety Administration). 1997. Relationship of Vehicle Weight to Fatality and Injury Risk in Model Year 1985–93 Passenger Cars and Light Trucks. NHTSA Summary Report, DOT HS 808 569. Springfield, Va.: National Technical Information Services.

Nivola, P.S., and R.W. Crandall. 1995. The Extra Mile. Washington, D.C.: The Brookings Institution.

NRC (National Research Council). 1992. Automotive Fuel Economy: How Far Should We Go? Washington, D.C.: National Academy Press.

NRC. 2001. Climate Change Science: An Analysis of Some Key Questions. Washington, D.C.: National Academy Press.

Plotkin, S. 2001. European and Japanese Initiatives to Boost Automotive Fuel Economy: What They Are, Their Prospects for Success, Their Usefulness As a Guide for U.S. Actions. Presentation to the 80th Annual Meeting, Transportation Research Board, National Research Council, Washington, D.C., January 7–11.

Tatom, J.A. 1993. "Are There Useful Lessons from the 1990–91 Oil Price Shock?" *The Energy Journal* 14(4):129–150.

United Auto Workers. 2000. Labor Organization. Annual Report for Fiscal Year Ending December 31, 2000. Form LM-2, December.

3

Technologies for Improving the Fuel Economy of Passenger Cars and Light-Duty Trucks

This chapter examines a variety of technologies that could be applied to improve the fuel economy of future passenger vehicles. It assesses their fuel economy potential, recognizing the constraints imposed by vehicle performance, functionality, safety, cost, and exhaust emissions regulations.

The committee reviewed many sources of information related to fuel economy-improving technologies and their associated costs, including presentations at public meetings and available studies and reports. It also met with automotive manufacturers and suppliers and used consultants to provide additional technical and cost information (EEA, 2001; Sierra Research, 2001). Within the time constraints of this study, the committee used its expertise and engineering judgment, supplemented by the sources of information identified above, to derive its own estimates of the potential for fuel economy improvement and the associated range of costs.

In addition, after the prepublication copy of the report was released in July 2001, the committee reexamined its technical analysis. Representatives of industry and other groups involved in fuel efficiency analysis were invited to critique the committee's methodology and results. Several minor errors discovered during this reexamination have been corrected in this chapter, and the discussion of the methodology and results has been clarified. The reexamination is presented in Appendix F.

FUEL ECONOMY OVERVIEW

To understand how the fuel economy of passenger vehicles can be increased, one must consider the vehicle as a system. High fuel economy is only one of many vehicle attributes that may be desirable to consumers. Vehicle performance, handling, safety, comfort, reliability, passenger- and load-carrying capacity, size, styling, quietness, and costs are also important features. Governmental regulations require vehicles to meet increasingly stringent requirements, such as reduced exhaust emissions and enhanced safety features. Ultimately these requirements influence final vehicle design, technology content, and—the subject of this report—fuel economy. Manufacturers must assess trade-offs among these sometimes-conflicting characteristics to produce vehicles that consumers find appealing and affordable.

Engines that burn gasoline or diesel fuel propel almost all passenger cars and light-duty trucks. About two-thirds of the available energy in the fuel is rejected as heat in the exhaust and coolant or frictional losses.[1] The remainder is transformed into mechanical energy, or work. Some of the work is used to overcome frictional losses in the transmission and other parts of the drive train and to operate the vehicle accessories (air conditioning, alternator/generator, and so on). In addition, standby losses occur to overcome engine friction and cooling when the engine is idling or the vehicle is decelerating.

As a result, only about 12 to 20 percent of the original energy contained in the fuel is actually used to propel the vehicle. This propulsion energy overcomes (1) inertia (weight) when accelerating or climbing hills, (2) the resistance of the air to the vehicle motion (aerodynamic drag), and (3) the rolling resistance of the tires on the road. Consequently, there are two general ways to reduce vehicle fuel consumption: (1) increase the overall efficiency of the powertrain (engine, transmission, final drive) in order to deliver more work from the fuel consumed or (2) reduce the required work (weight, aerodynamics, rolling resis-

---

[1]Theoretically gasoline or diesel engines (and fuel cells) can convert all of the fuel energy into useful work. In practice, because of heat transfer, friction, type of load control, accessories required for engine operation, passenger comfort, etc., the fraction used to propel the vehicle varies from as low as zero (at idling) to as high as 40 to 50 percent for an efficient diesel engine (gasoline engines are less efficient). Further losses occur in the drive train. As a result, the average fraction of the fuel converted to work to propel the vehicle over typical varying-load operation is about 20 percent of the fuel energy.

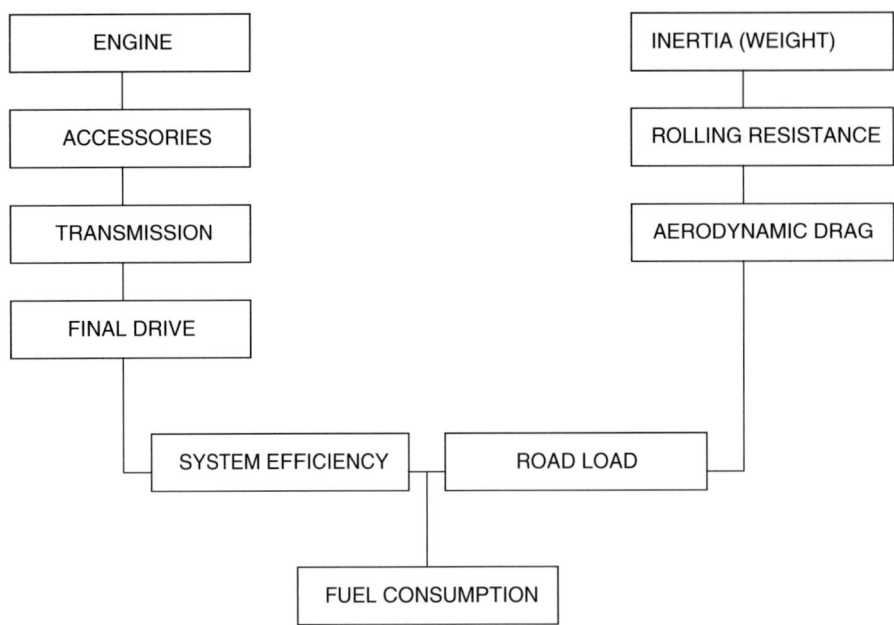

FIGURE 3-1 Energy use in vehicles. SOURCE: Adapted from Riley (1994).

tance, and accessory load) to propel the vehicle. These concepts are illustrated in Figures 3-1 and 3-2. Regenerative braking and shutting the engine off during idling also save energy, as discussed in the section on hybrid electric vehicles, below.

Vehicle fuel economy currently is determined according to procedures established by the Environmental Protection Agency (EPA). Vehicles are driven on a dynamometer in a controlled laboratory (in order to eliminate weather and road variables.)[2] Both city and highway driving are simulated. The city test is a 7.5-mile trip lasting 23 minutes with 18 stops, at an average speed of about 20 miles per hour (mph). About 4 minutes are spent idling (as at a traffic light), and a short freeway segment is included. The vehicle begins the test after being parked overnight at about 72°F (22°C) (cold soak). The highway test is a 10-mile trip with an average speed of about 48 mph. The test is initiated with a warmed-up vehicle (following the city test) and is conducted with no stops and very little idling. The basis for compliance with CAFE (and comparison of the technologies below) is the current EPA Federal Test Procedure (FTP-75) with city, highway, and combined (55 percent city/45 percent highway) ratings in miles per gallon (mpg) (CFR, 2000).

During city driving, conditions such as acceleration, engine loading, and time spent braking or at idle are continually changing across a wide range of conditions. These variations result in wide swings in fuel consumption. Inertial loads and rolling resistance (both directly related to weight) combined account for over 80 percent of the work required to move the vehicle over the city cycle, but less for the highway cycle. A reduction in vehicle weight (mass) therefore has a very significant effect on fuel consumption in city driving. This strong dependence on total vehicle weight explains why fuel consumption for the new vehicle fleet correlates linearly with vehicle curb weight, as shown in Attachment 3A.

Weight reduction provides an effective method to reduce fuel consumption of cars and trucks and is an important goal for the government-industry program Partnership for a New Generation of Vehicles (PNGV). Reducing the required propulsion work reduces the load required from the engine, allowing the use of a smaller engine for the same performance. In the search for lightweight materials, PNGV has focused on materials substantially lighter than the steel used in most current vehicles. Components and body structure fabricated from aluminum, glass-fiber-reinforced polymer composites (GFRP), and carbon-fiber-reinforced polymer composites (CFRP), including hybrid structures, are being investigated (NRC, 2000a).

Reducing vehicle weight without reducing practical space for passengers and cargo involves three strategies: (1) substitution of lighter-weight materials without compromising

---

[2]Aerodynamic drag is accounted for in the results by incorporating coastdown data from other tests. Nevertheless, there are significant differences between the mileage tests and real-life driving. For example, the dynamometer is connected to only one pair of tires, but on the road, all tires are rolling. Most drivers experience lower fuel economy than suggested by EPA's results. It should be noted that the test driving cycles were derived from traffic pattern observations made many years ago, which may not be representative now. A review of the validity of the test cycles for today's traffic patterns would seem appropriate.

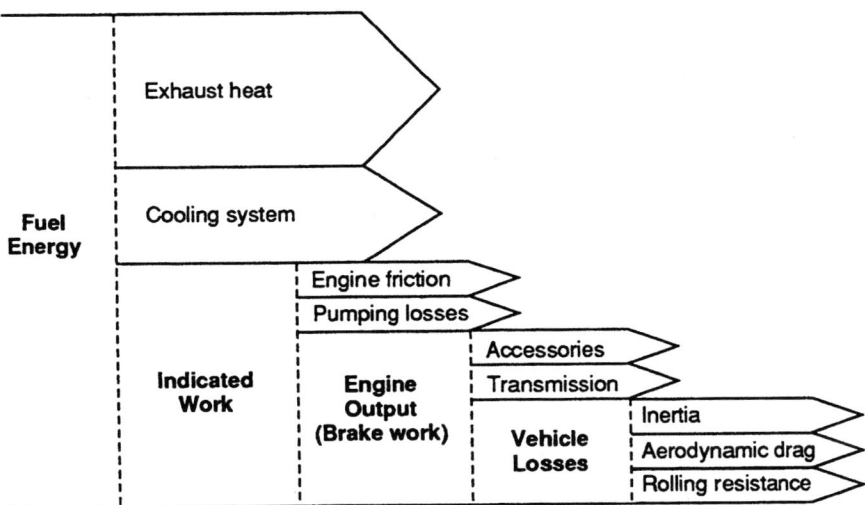

FIGURE 3-2 Where the energy in the fuel goes (proportions vary with vehicle design, type of engine, and operating conditions). SOURCE: NRC (1992).

structural strength (e.g., aluminum or plastic for steel); (2) improvement of packaging efficiency, that is, redesign of the drive train or interior space to eliminate wasted space; and (3) technological change that eliminates equipment or reduces its size. Design efficiency and effectiveness can also result in lighter vehicles using the same materials and the same space for passengers and cargo.

Automotive manufacturers must optimize the vehicle and its powertrain to meet the sometimes-conflicting demands of customer-desired performance, fuel economy goals, emissions standards, safety requirements, and vehicle cost within the broad range of operating conditions under which the vehicle will be used. This necessitates a vehicle systems analysis. Vehicle designs trade off styling features, passenger value, trunk space (or exterior cargo space for pickups), and utility. These trade-offs will likewise influence vehicle weight, frontal area, drag coefficients, and power train packaging, for example. These features, together with engine performance, torque curve, transmission characteristics, control system calibration, noise control measures, suspension characteristics, and many other factors, will define the drivability, customer acceptance, and marketability of the vehicle.

Technology changes modify the system and hence have complex effects that are difficult to capture and analyze. It is usually possible, however, to estimate the impacts of specific technologies in terms of a percentage savings in fuel consumption for a typical vehicle without a full examination of all the system-level effects. Such a comparative approach is used in this chapter.[3]

Although CAFE standards and EPA fuel economy ratings are defined in the now-familiar term miles per gallon (mpg), additional assessment parameters have been identified to assist in the evaluation process, including fuel consumption in gallons per 100 miles; load-specific fuel consumption (LSFC) in gallons per ton (of cargo plus passengers) per 100 miles; and weight-specific fuel consumption (WSFC) in gallons per vehicle weight per 100 miles. Attachment 3A further explains why these parameters are meaningful engineering relationships by which to judge fuel economy and the efficiency of moving the vehicle and its intended payload over the EPA cycle.

Figure 3-3 shows the actual energy efficiency of vehicles of different weights. For both city and highway cycles, the fuel consumed per ton of weight and per 100 miles is plotted against the weight of the vehicle. Normalizing the fuel consumption (dividing by weight) yields an efficiency factor (in an engineering sense), which is particularly useful in comparing fuel savings opportunities. It is also useful that the weight-adjusted or normalized values can be reasonably approximated by a horizontal line. Points above the line represent vehicles with lower-than-average efficiency, which require more than the average amount of fuel to move a vehicle of a given weight over the EPA certification cycle. In principle, given sufficient lead time and business incentives (economic or regulatory), the vehicles above the line could be improved to the level of those below the line, within the limits of customer-desired performance and vehicle utility features. As an example, a 4,000-lb vehicle, in principle, could drop from 2 gallons per ton-100 miles (25 mpg) on the highway cycle to 1.4 (35.7 mpg) using technologies discussed later in this chapter. However, although larger, heavier vehicles have greater fuel consumption than smaller, lighter vehicles, their energy efficiencies in moving the vehicle mass (weight) are very similar. These data also suggest that the potential exists to improve fuel consumption in future vehicles. However, changing conditions such as safety

---

[3]Further explanation of the methodology is provided in Appendix F.

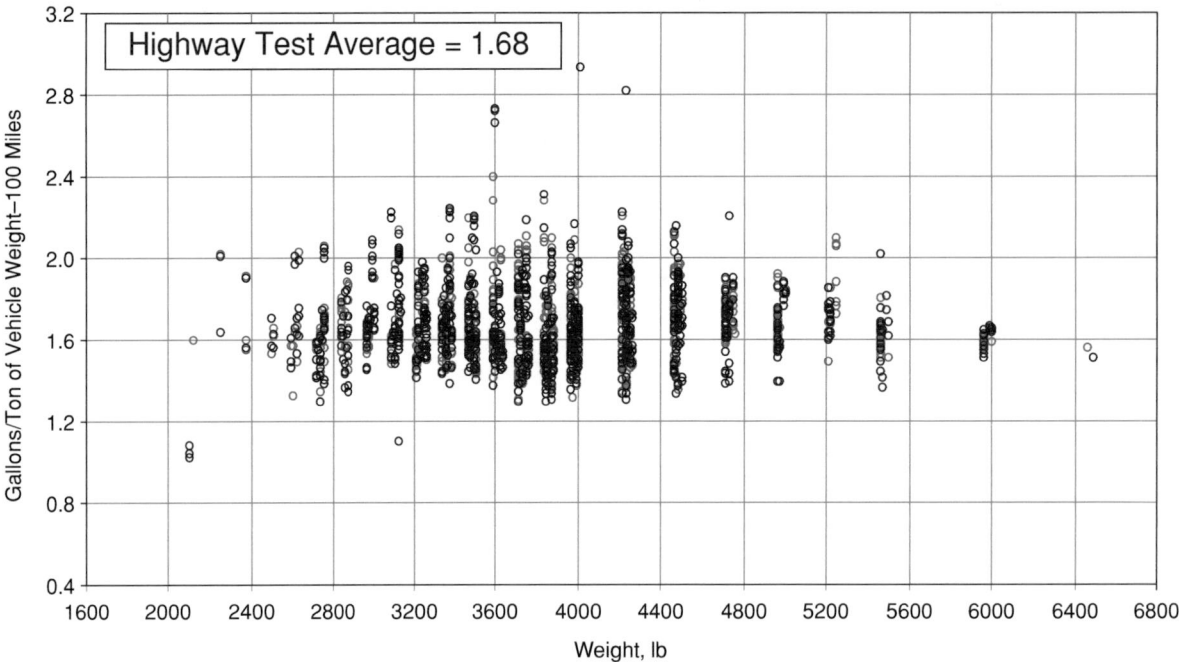

FIGURE 3-3 EPA data for fuel economy for MY 2000 and 2001 cars and light trucks. SOURCE: EPA, available online at <http://www.epa.gov/otaq/mpg.htm>.

regulations, exhaust emission standards, consumer preferences, and consumer-acceptable costs must be traded off. The remainder of this chapter attempts to outline this complex relationship.

Future Exhaust Emission and Fuel Composition Standards

New environmental regulations will have a significant impact on certain technologies that have demonstrated the potential for significantly improving fuel economy. In particular, the possible introduction of diesel engines and lean-burn, direct-injection gasoline engines will be affected. Oxides of nitrogen (NO_x) and particulate matter (PM) standards are particularly stringent (NRC, 2000b) compared, for instance, with current and future standards in Europe, where diesel and lean-burn gasoline have significant market penetration.

The Clean Air Act Amendments of 1990 imposed new federal regulations on automotive emissions and authorized EPA to determine the need for and cost and feasibility of additional standards. EPA made this determination and will initiate so-called Tier 2 regulations, phasing them in over model years 2004 to 2009. The Tier 2 standards are very complex and will not be addressed here in detail. However, certain key features will be mentioned as they impact potential fuel economy gains. Unlike current emission standards, Tier 2 standards will vary depending on vehicle type rather than weight class. Interim phase-in schedules and durability requirements (the life expectancy that must be demonstrated for emission control systems) also vary by weight.

Emissions from large sport utility vehicles (SUVs) and passenger vans weighing between 8,500 and 10,000 lb gross vehicle weight (GVW), which are currently exempt under Tier 1, will be regulated under Tier 2 standards. However, pickup trucks in this same weight range (NRC, 2000b) will not. EPA has also promulgated regulations to reduce sulfur in gasoline to 30 parts per million (ppm) (EPA, 1999) and in diesel fuel to 15 ppm (EPA, 2000b).

Tier 2 includes a "bin" system that allows manufacturers to average emissions across the fleet of vehicles they sell each year, unlike the current system that requires each vehicle to meet the same emissions standard. Vehicles certified in a particular bin must meet all of the individual emission standards (NO_x, nonmethane organic gases, CO, formaldehyde, PM) for that bin. In addition, the average NO_x emissions level of the entire fleet sold by a manufacturer will have to meet the standard of 0.07 g/mile. During the phase-in period, 10 bins will be allowed, but after 2009, the 2 most lenient bins will be dropped.

EPA has communicated its belief that the combination of bins, averaging, and a phase-in period could allow the introduction of new diesel and other high-efficiency engine technology. However, the high development and production costs for such engines, combined with the high uncertainty of meeting the ultimately very low NO_x and PM standards, even with the reduced sulfur level in diesel fuel (15 ppm) that will be available in 2006 (EPA, 2000b), has delayed production decisions. In general, the committee believes that the Tier 2 NO_x and PM standards will inhibit, or possibly preclude, the introduction of diesels into vehicles under 8,500 lb unless cost-effective, reliable, and regulatory-compliant exhaust gas aftertreatment technology develops rapidly. A key challenge is the development of emission control systems that can be certified for a 120,000-mile lifetime.

In theory, the bin system will allow diesels to penetrate the light-duty vehicle market, but manufacturers must still meet the stringent fleet average standard. For example, for every vehicle in bin 8 (0.2 g/mile NO_x), approximately seven vehicles in bin 3 (0.03 g/mile) would have to be sold in order to meet the 0.07 g/mile fleet-average NO_x standard.

These same factors have caused the committee to conclude that major market penetration of gasoline direct-injection engines that operate under lean-burn combustion, which is another emerging technology for improving fuel economy, is unlikely without major emissions-control advancements.

California's exhaust emission requirements—super ultralow emission vehicle (SULEV) and partial zero emission vehicle (PZEV)—are also extremely challenging for the introduction of diesel engines. In particular, the California Air Resources Board (CARB) has classified PM emissions from diesel-fueled engines as a toxic air contaminant (CARB, 1998). (Substances classified as toxic are required to be controlled.)

TECHNOLOGIES FOR BETTER FUEL ECONOMY

The 1992 NRC report outlined various automotive technologies that were either entering production at the time or were considered as emerging, based on their potential and production intent (NRC, 1992). Since then, many regulatory and economic conditions have changed. In addition, automotive technology has continued to advance, especially in microelectronics, mechatronics, sensors, control systems, and manufacturing processes. Many of the technologies identified in the 1992 report as proven or emerging have already entered production.

The committee conducted an updated assessment of various technologies that have potential for improving fuel economy in light-duty vehicles. This assessment takes into account not only the benefits and costs of applying the technologies, but also changes in the economic and regulatory conditions, anticipated exhaust emission regulations, predicted trends in fuel prices, and reported customer preferences.

The technologies reviewed here are already in use in some vehicles or are likely to be introduced in European and Japanese vehicles within 15 years. They are discussed below under three general headings: engine technologies, transmission technologies, and vehicle technologies. They are listed in general order of ease of implementation or maturity of the technology (characterized as "production intent" or "emerging"). The committee concludes its assessment of potential technologies with some detailed discussion of the current and future generations of hybrid vehicles and fuel-cell power sources.

For each technology assessed, the committee estimated not only the incremental percentage improvement in fuel consumption (which can be converted to fuel economy in miles per gallon [mpg] to allow comparison with current EPA mileage ratings) but also the incremental cost that applying the technology would add to the retail price of a vehicle. The next subsection of this chapter, "Technologies Assessed," reviews the technologies and their general benefits and challenges.

After that, the section "Estimating Potential Fuel Economy Gains and Costs" presents estimates of the fuel consumption benefits and associated retail costs of applying combinations of these technologies in 10 classes of production vehicles. For each class of vehicle, the committee hypothesizes three exemplary technology paths (technology scenarios leading to successively greater improvements in fuel consumption and greater cost).

Technologies Assessed

The engine, transmission, and vehicle technologies discussed in this section are all considered likely to be available within the next 15 years. Some (called "production intent" in this discussion) are already available, are well known to manufacturers and their suppliers, and could be incorporated in vehicles once a decision is reached to use them. Others (called "emerging" in this discussion) are generally beyond the research phase and are under development. They are sufficiently well understood that they should be available within 10 to 15 years.

Engine Technologies

The engine technologies discussed here improve the energy efficiency of engines by reducing friction and other mechanical losses or by improving the processing and combustion of fuel and air.

Production-Intent Engine Technologies The engine technologies discussed here could be readily applied to production vehicles once a decision is made to proceed, although various constraints may limit the rate at which they penetrate the new vehicle fleet:

- *Engine friction and other mechanical/hydrodynamic loss reduction.* Continued improvement in engine component and system design, development, and computer-aided engineering (CAE) tools offers the potential for continued reductions of component weight and thermal management and hydrodynamic systems that improve overall brake-specific efficiency. An im-

provement in fuel consumption of 1 to 5 percent is considered possible, depending on the state of the baseline engine.
- *Application of advanced, low-friction lubricants.* The use of low-friction, multiviscosity engine oils and transmission fluids has demonstrated the potential to reduce fuel consumption by about 1 percent, compared with conventional lubricants.
- *Multivalve, overhead camshaft valve trains.* The application of single and double overhead cam designs, with two, three, or four valves per cylinder, offers the potential for reduced frictional losses (reduced mass and roller followers), higher specific power (hp/liter), engine downsizing, somewhat increased compression ratios, and reduced pumping losses. Depending on the particular application and the trade-offs between valve number, cost, and cam configuration (single overhead cam [SOHC] or double overhead cam [DOHC]), improvements in fuel consumption of 2 to 5 percent are possible, at constant performance, including engine downsizing (Chon and Heywood, 2000). However, market trends have many times shown the use of these concepts to gain performance at constant displacement, so that overall improvements in fuel consumption may be less.
- *Variable valve timing (VVT).* Variation in the cam phasing of intake valves has gained increasing market penetration, with an associated reduction in production cost. Earlier opening under low-load conditions reduces pumping work. Under high-load, high-speed conditions, variations in cam phasing can improve volumetric efficiency (breathing) and help control residual gases, for improved power. Improvements in fuel consumption of 2 to 3 percent are possible through this technology (Chon and Heywood, 2000; Leone et al., 1996).
- *Variable valve lift and timing (VVLT).* Additional benefits in air/fuel mixing, reduction in pumping losses, and further increases in volumetric efficiency can be gained through varying timing and valve lift (staged or continuous). Depending upon the type of timing and lift control, additional reductions in fuel consumption of 1 to 2 percent, above cam phasing only, are possible (Pierik and Burkhanrd, 2000), or about 5 to 10 percent compared to two-valve engines (including downsizing with constant performance).
- *Cylinder deactivation.* An additional feature that can be added to variable valve lift mechanisms is to allow the valves of selected cylinders to remain closed, with the port fuel injection interrupted. Currently, this technology is applied to rather large engines (>4.0 liter) in V8 and V12 configurations. This approach, which is sometimes referred to as a variable displacement engine, creates an "air spring" within the cylinder. Although both frictional and thermodynamic losses occur, they are more than offset by the increased load and reduced specific fuel consumption of the remaining cylinders. However, engine transient performance, idle quality, noise, and vibration can limit efficiency gains and must be addressed. Improvements in fuel consumption in the range of 3 to 6 percent are possible, even given that reductions in throttling losses associated with higher load factors over the operating cycle cannot be double counted.
- *Engine accessory improvement.* As engine load and speed ranges continue to advance, many engine accessories such as lubrication and cooling systems and power steering pumps are being optimized for reductions in energy consumption and improved matching of functionality over the operating range. The evolution of higher-voltage (i.e., 42 V) powertrain and vehicle electrical systems will facilitate the cost-efficient applications of such components and systems. Improvements in fuel consumption of about 1 to 2 percent are possible with such technologies.
- *Engine downsizing and supercharging.* Additional improvements in fuel consumption can be gained by reducing engine displacement and increasing specific power (while maintaining equal performance) by boosting the engine (turbocharger or mechanical supercharger). Degraded transient performance (turbo-lag) typically associated with turbochargers can be significantly offset by incorporating variable geometry turbines or mechanical (positive displacement) superchargers. Additional modifications for transmission matching, aftertreatment system warm-up, and other factors that can degrade exhaust emissions control must also be considered. Improvements in fuel consumption of 5 to 7 percent are considered possible with this approach, at equivalent vehicle performance (Ecker, 2000). However, when this concept is combined with multivalve technology, total improvements of about 10 percent are possible compared with a two-valve engine baseline.

Emerging Engine Technologies The following engine technologies are considered emerging for passenger car and light-duty truck applications. Significant market penetration in the United States is likely to take 5 to 10 years. Some of them are already in production elsewhere (in Japan or Europe), where they may benefit from high fuel taxes, government incentives for particular engine types or displacements, and more lenient exhaust emission or vehicle safety standards. The discussion that follows outlines not only the benefits but also the technical challenges or economic hurdles for each technology.

- *Intake valve throttling (IVT).* Advances in microprocessor technology, feedback control, electromechanical actuation, sensor technology, and materials continue to accelerate. As a result, electromechanical IVT is advancing to the point where BMW has announced the introduction of its so-called Valvetronic concept. When multipoint fuel injection is used, both the lift and timing of the intake valves can be controlled to maintain the correct air/fuel ratio without a throttle plate. This has the potential to essentially eliminate the pumping losses across the normal butterfly throttle valve. Also important is the potential to use conventional three-way-catalyst (TWC) aftertreatment technology and incorporate cylinder deactivation. However, significant cost and complexity in actuation, electronic control, and system calibration are to be expected. Improvements in fuel consumption of an additional 3 to 6 percent above VVLT are possible with this technology. Compared with two-valve engines, total system improvements may approach 6 to 12 percent.

- *Camless valve actuation (CVA)*. A further evolution of fast-acting, completely variable valve timing (not limited by the lift curve of a camshaft) is represented by electromechanical solenoid-controlled, spring-mass valve (EMV) systems (Siemens, BMW, FEV) and high-pressure hydraulic-actuated valves with high-speed, digital control valve technology (Ford, Navistar). In addition to reducing pumping losses, this technology facilitates intake port and cylinder deactivation and allows the use of conventional TWC aftertreatment. Technical challenges in the past for EMV have been to minimize energy consumption and achieve a soft landing of the valve against the seat during idle and low-speed, low-load operation, for acceptable noise levels. These issues appear to be solved through advances in sensor and electromagnetic technologies. EMV systems are expected to see limited production within 5 to 7 years. Improvements in fuel consumption of 5 to 10 percent relative to VVTL are possible with this technology. Compared with fixed-timing, four-valve engines, total system improvements of 15 percent or more have been demonstrated (Pischinger et al., 2000).
- *Variable compression ratio (VCR) engines*. Current production engines are typically limited in compression ratio (CR) to about 10:1 to 10.5:1 with the use of high-octane fuel, owing to knocking under high load. However, significant improvements in fuel consumption could be gained with higher CR under normal driving cycles. Many different VCR approaches that allow improved efficiency under low load with high CR (13-14:1) and sufficient knock tolerance under full load with lower CR (~8:1) are under development. Saab appears to have the most advanced VCR prototypes. Automakers, suppliers, and R&D organizations are currently exploring many other approaches that are applicable to both inline and Vee engine configurations. Several of these are expected to enter production within 10 years. Compared with a conventional four-valve VVT engine, improvements in fuel consumption of 2 to 6 percent are possible (Wirbeleit et al., 1990). The combination of VCR with a supercharged, downsized engine is likely to be effective, giving the maximum advantage of both systems and reducing total fuel consumption, at constant performance, by 10 to 15 percent. However, the potential complexity of the hardware, system durability, control system development, and cost must be traded off for production applications.

Many additional engine technologies with good potential for improved fuel consumption are the subject of R&D. Others are currently offered in markets with higher fuel prices (due to higher taxes) or exhaust emission standards more lenient than the upcoming federal Tier 2 emission standards (or California's SULEV standards, set to begin in model year 2004). A brief summary of these technologies is presented below, including reference to the areas of uncertainty and the need for further development.

- *Direct-injection (DI) gasoline engines*. Stratified-charge gasoline engines burning in a lean mode (when more air is present than required to burn the fuel) offer improved thermodynamic efficiency. However, the technology faces potential problems in controlling particulate emissions and NO_x. Trade-offs between the maximum operating range under lean conditions versus stoichiometric operation (when the exact amount of air needed to burn the fuel is present) with early injection must be developed. Although lean-burn DI engines of the type offered in Europe could improve fuel consumption by more than 10 percent, NO_x-control requirements that necessitate stoichiometric operation and the use of TWCs limit the potential fuel consumption improvement to between 4 and 6 percent (Zhao and Lai, 1997).
- *Direct-injection diesel engines*. The application of small (1.7- to 4.0-liter), high-speed (4,500-rpm), turbocharged, direct-injection diesel engines has seen tremendous expansion in passenger cars and light-duty trucks in Europe. Increasing power densities (>70 hp/liter), achieved through the application of advanced, high-pressure, common-rail fuel injection systems; variable geometry turbochargers; and advances in noise, vibration, and harshness (NVH) control technologies, combined with high-efficiency, lean-burn combustion systems and practically smokeless and odorless emissions, have greatly improved customer acceptance in Europe. The high low-speed torque and relatively flat torque curve also offer significant drivability improvements. Fuel consumption improve-

ments of 30 to 40 percent or more are possible compared with conventional two-valve gasoline engines. The challenges, which inhibit widespread introduction in the United States, include meeting strict NO_x and particulate emission standards for Tier 2 and SULEV; much higher engine and vehicle purchase price ($2,000 to $3,000) than conventional gasoline engines; and uncertain U.S. customer acceptance. The creation of NO_x and particulate emissions is exacerbated by the stratification present in the fuel/air mixture resulting from in-cylinder injection. R&D activities continue on emission control through advanced combustion and fuel injection concepts, fuel composition (low sulfur), aftertreatment technologies (selective catalytic reduction [SCR], NO_x traps, particulate filters), and control of noise and vibration. Although DI diesel engines are offered in some trucks over 8,500 lb and are offered by one manufacturer (VW) for passenger cars under current Tier 1 emissions standards, wide use in sport utility vehicles (SUVs) and light-duty trucks has not occurred, and the ability of this technology to comply with the upcoming Tier 2 and SULEV standards is highly uncertain.

Transmission Technologies

The second group of technologies assessed by the committee involves improvements in the efficiency with which power is transmitted from the engine to the driveshaft or axle.

Production-intent Transmission Technologies Over the past 10 years, transmission technologies have been evolving toward increasing electronic control, adapting torque converter lock-up, four- and five-speed automatics (from three- and four-speed), and various versions of all-wheel drive (AWD) or four-wheel drive (4WD) and traction control, ranging from continuous, traction-controlled AWD to automatic 2WD-4WD traction control in some SUVs.

- *Five-speed automatic transmission.* A five-speed automatic transmission permits the engine to operate in its most efficient range more of the time than does a four-speed transmission. A fuel consumption improvement of 2 to 3 percent is possible, at constant vehicle performance, relative to a four-speed automatic.
- *Continuously variable transmission (CVT).* Several versions of continuously variable transmissions are offered in production in Europe and Japan and a few in the United States (by Honda and Toyota). Historically, these transmission types have used belts or chains of some kind to vary speed ratios across two variable-diameter pulleys. The major production units utilize compression belts (VanDorne) or tension chains. Other approaches are also being pursued for future production. Depending on the type of CVT and the power/speed range of the engine, this technology can improve fuel consumption by about 4 to 8 percent. However, production cost, torque limitations, and customer acceptance of the system's operational characteristics must be addressed.

Emerging Transmission Technologies Automotive manufacturers continue to seek ways to reduce the mechanical (frictional and hydrodynamic) losses of transmissions and improve their mating with engines. The various types of hybrid vehicles will also involve changes in conventional transmissions. These emerging technologies are likely to be available in the latter part of the current decade.

- *Automatic transmission with aggressive shift logic.* Shift schedules, logic, and control of torque transfer can significantly affect perceived shift quality. Advanced work on methods to reduce losses associated with torque converters or torque dropout is being pursued. It is estimated that a 1 to 3 percent improvement in fuel consumption can be obtained through such measures. However, these will be highly affected by customer perception in the United States and may require quite some time for significant acceptance.
- *Six-speed automatic transmission.* Advanced six-speed automatic transmissions can approach the performance of CVT transmissions without limitations in the ability to transmit torque. An additional improvement of 1 to 2 percent in fuel consumption is possible, compared to a five-speed automatic. Based on their higher cost and control complexity, such transmissions will probably see only limited introduction—namely, in high-end luxury or performance vehicles.
- *Automatic shift/manual transmission (ASM/AMT).* In the continuing quest to reduce mechanical losses, manufacturers are developing new generations of automatic transmissions that eliminate the hydraulic torque converter and its associated pump, replacing it with electronically controlled clutch mechanisms. This approach offers two basic possibilities: The torque from different gear sets can be intermittently interrupted (as in a conventional manual transmission) through the use of a single electronically controlled clutch; or the torque can be continuously controlled, without dropout, through the use of two electronically controlled clutches. Improvements in fuel consumption of 3 to 5 percent over a conventional four-speed automatic transmission with hydraulic torque converter are possible. However, increased cost, control system complexity, durability, and realizable fuel consumption gain versus acceptable shift quality for U.S. customers must be addressed.

- *Advanced CVT.* Continued advances in methods for high-efficiency, high-torque transfer capability of CVTs are being pursued. New versions of CVTs that will soon enter production incorporate toroidal friction elements or cone-and-ring assemblies with varying diameters. However, these versions also have trade-offs of torque capability vs. frictional losses. These next-generation transmissions have the potential to improve fuel consumption by about 0 to 2 percent (relative to current CVTs), with higher torque capabilities for broader market penetration. However, production cost, system efficiency, and customer acceptance of the powertrain operational characteristics must still be addressed.

Vehicle Technologies

By reducing drag, rolling resistance, and weight, the fuel consumption of vehicles could, in principle, be cut rather sharply in the relatively near term. Manufacturers, however, would quickly run into serious trade-offs with performance, carrying capacity, and safety. Also to be considered are novel vehicle concepts such as hybrid electrics, powered by various combinations of internal combustion engines and batteries or fuel cells. The following discussion reviews both production-intent and emerging vehicle technologies.

Production-Intent Vehicle Technologies The following fuel consumption measures are deemed available in the near term (almost immediately after a decision to use them is made):

- *Aerodynamic drag reduction on vehicle designs.* This improvement can be very cost-effective if incorporated during vehicle development or upgrades. However, vehicle styling and crashworthiness have significant influences on the ultimate levels that can be achieved. For a 10 percent reduction in aerodynamic drag, an improvement in fuel consumption of 1 to 2 percent can be achieved. As drag coefficients proceed below about 0.30, however, the design flexibility becomes limited and the relative cost of the vehicle can increase dramatically. Substituting video minicameras for sideview mirrors (e.g., as is being done for the PNGV concept vehicles) would be advantageous but would necessitate a change in safety regulations (NRC, 2000a).
- *Rolling resistance.* Continued advances in tire and wheel technologies are directed toward reducing rolling resistance without compromising handling, comfort, or braking. Improvements of about 1 to 1.5 percent are considered possible. The impacts on performance, comfort, durability, and safety must be evaluated, however.
- *Vehicle weight reduction.* Reducing vehicle weight while maintaining acceptable safety is a difficult balance to define. While most manufacturers believe that some reduction in vehicle weight can be accomplished without a measurable influence on in-use safety, debate continues on how much weight can be reduced without compromising crush space, by using lighter-weight materials and better, more crashworthy designs.

Emerging Vehicle Technologies Several advanced vehicle technologies are being considered for near-term production. Interest in these technologies has been fostered by the PNGV program. In addition, a wide variety of hybrid vehicle technologies are being explored for initial introduction within the next 5 to 10 years. This section reviews vehicle technologies that have been identified by the industry for introduction within the next 10 years.

- *Forty-two volt electrical system.* Most automotive manufacturers are planning a transition to 42-V electrical systems to support the continuing need for increased electrical power requirements for next-generation passenger vehicles. Higher voltage will reduce electrical losses and improve the efficiency of many onboard electrically powered systems. It will also allow new technologies such as electric power steering, which can be significantly more efficient than current technology. Fuel consumption reductions associated with the implementation and optimization of related systems are expected to range from 1 to 2 percent.
- *Integrated starter/generator (ISG).* Significant improvements in fuel consumption under real-world operating conditions can be gained by turning the engine off during idle, while operating the necessary accessories electrically (air conditioning presents a major challenge, however). ISG systems providing nearly instantaneous engine restart are now planned for production. Idle stop, under many conditions, is expected to achieve a 4 to 7 percent reduction in fuel consumption. Depending on the size and type of battery chosen, it is also possible to recover electrical energy through regenerative braking and subsequent launch assist using ISG technology. Doing so adds cost, complexity, and weight but could improve fuel consumption by a total of 5 to 10 percent.
- *Hybrid electric vehicles.* Hybrid electric vehicles of various types are in different stages: Some are starting to be introduced, others are in advanced stages of development, and still others are the focus of extensive research by nearly all the large automotive manufacturers. They include so-called "mild hybrids" (with regenerative braking, ISG, launch assist, and minimal battery storage); "parallel hybrids" (with the engine

powering either or both a mechanical drive train and an electric motor/generator serving as additional propulsion to recharge the battery); and "series hybrids" (in which the engine does not drive the wheels but always drives an electric motor/generator to propel the vehicle, recharge the battery, or perform both functions simultaneously).

The method and extent of hybridization depends on the vehicle type, its anticipated use, accessory package, type of battery, and other considerations. The anticipated improvements in fuel consumption can therefore vary, from about 15 percent for certain mild hybrids to about 30 percent for parallel hybrids. In general, series hybrids are not yet intended for even limited production, owing to the relatively poor performance of electric power propulsion and the low efficiencies of current battery systems compared with mechanical drive systems. The varying complexity of the different hybrid types is reflected in large variations in incremental cost. The cost premium of today's limited-production mild hybrids is predicted to be $3,000 to $5,000 when they reach production volumes over 100,000 units per year. For fully parallel systems, which operate for significant periods entirely on the electrical drive, especially in city driving, the cost premium can escalate to $7,500 or more. In addition to offering significant gains in fuel consumption, these vehicles have the potential for beneficial impacts on air quality. Further information on hybrids is provided in a separate section below.

- *Fuel-cell hybrid electric vehicles.* The most advanced emerging vehicle technology currently under research and development substitutes an electrochemical fuel cell for the internal combustion engine. In proton exchange membrane (PEM) fuel cells, hydrogen and oxygen react to produce electricity and water. Since gaseous hydrogen is difficult to store with reasonable energy density, many manufacturers are pursuing the decomposition of a liquid fuel (either methanol or gasoline) as a source of hydrogen, depending on corporate perceptions of future fuel availability. State-of-the-art fuel cell systems demonstrate the potential for long-term viability: They could realize high electrochemical energy conversion efficiencies and very low local exhaust emissions, depending upon the type of fuel chosen and the associated reformation process to produce hydrogen. However, the presence of sulfur in gasoline could pose a significant problem—PEM poisoning. Owing to its high potential for reducing fuel consumption, this emerging technology is receiving substantial R&D funding. However, most researchers and automotive manufacturers believe that successful commercial application of fuel cells for passenger vehicles is at least 10 to 15 years away. Further information on fuel cells is provided in a separate section later in this chapter.

With the exception of fuel cells and series hybrids, the technologies reviewed above are all currently in the production, product planning, or continued development stage, or are planned for introduction in Europe or Japan. The feasibility of production is therefore well known, as are the estimated production costs. However, given constraints on price imposed by competitive pressures in the U.S. market, only certain technologies are considered practical or cost effective.

As noted earlier, the exhaust emission standards in the United States (Tier 2 and SULEV) make the introduction of some high-fuel-economy technologies, such as lean-burn, direct-injection gasoline or high-speed DI diesel engines, uncertain. For these technologies to be viable, low-sulfur fuel must be available and particulate traps and NO_x emissions controls (lean NO_x catalyst, NO_x trap, SCR) must be developed. Therefore, current powertrain strategies for gasoline-powered engines use mainly stoichiometric air/fuel mixtures, for which three-way-catalyst aftertreatment is effective enough to meet future emission standards.

ESTIMATING POTENTIAL FUEL ECONOMY GAINS AND COSTS

To predict the costs associated with achieving improvements in fuel consumption, it is necessary to assess applications of the committee's list of technologies in production vehicles of different types. The committee estimated the incremental fuel consumption benefits and the incremental costs of technologies that may be applicable to actual vehicles of different classes and intended uses. The committee has hypothesized three successively more aggressive (and costly) product development paths for each of 10 vehicle classes to show how economic and regulatory conditions may affect fuel economy:

- *Path 1.* This path assumes likely market-responsive or competition-driven advances in fuel economy using production-intent technology that may be possible under current economic (fuel price) and regulatory (CAFE, Tier 2, SULEV) conditions and could be introduced within the next 10 years. It holds vehicle performance constant and assumes a 5 percent increase in vehicle weight associated with safety-enhancing features.
- *Path 2.* This path assumes more aggressive advances in fuel economy that employ more costly production-intent technologies but that are technically feasible for introduction within the next 10 years if economic and/or regulatory conditions justify their use. It also maintains constant vehicle performance and assumes a 5

percent increase in vehicle weight associated with safety-enhancing features.
- *Path 3.* This path assumes even greater fuel economy gains, which would necessitate the introduction of emerging technologies that have the potential for substantial market penetration within 10 to 15 years. These emerging technologies require further development in critical aspects of the total system prior to commercial introduction. However, their thermodynamic, mechanical, electrical, and control features are considered fundamentally sound. High-speed, direct-injection diesel engines, for instance, are achieving significant market penetration in Europe. However, strict exhaust emission standards in the United States necessitate significant efforts to develop combustion or exhaust aftertreatment systems before these engines can be considered for broad introduction.

For each product development path, the committee estimated the feasibility, potential incremental fuel consumption improvement, and incremental cost for 10 vehicle classes:

- Passenger cars: subcompact, compact, midsize, and large;
- Sport utility vehicles: small, midsize, and large; and
- Other light trucks: small pickup, large pickup, and minivan.

The three paths were estimated to represent vehicle development steps that would offer increasing levels of fuel economy gain (as incremental relative reductions in fuel consumption) at incrementally increasing cost. The committee has applied its engineering judgment in reducing the otherwise nearly infinite variations in vehicle design and technology that would be available to some characteristic examples. The approach presented here is intended to estimate the potential costs and fuel economy gains that are considered technically feasible but whose costs may or may not be recoverable, depending on external factors such as market competition, consumer demand, or government regulations.

The committee assembled cost data through meetings and interviews with representatives of automotive manufacturers and component and subsystem suppliers and through published references. Cost estimates provided by component manufacturers were multiplied by a factor of 1.4 to approximate the retail price equivalent (RPE) costs for vehicle manufacturers to account for other systems integration, overhead, marketing, profit, and warranty issues (EEA, 2001).

Experience with market competition has shown that the pricing of products can vary significantly, especially when the product is first introduced. Furthermore, marketing strategies and customer demand can greatly influence the RPE cost passed along to customers. Retail prices vary greatly, especially for components required to meet regulatory standards (such as catalytic converters, air bags, or seat belts).

The baseline fuel economies for these evolutionary cases are the lab results (uncorrected for on-road experience) on the 55/45 combined cycle for MY 1999 for each vehicle class (EPA, 2001a). Both the average fuel economy (in mpg) and the initial fuel consumption (in gallons/100 miles) are shown in Tables 3-1 through 3-3. The incremental improvements, however, were calculated as percentage reductions in fuel consumption (gallons per 100 miles). (The two measures should not be confused; a 20 percent decrease in fuel consumption, for example, from 5 gallons per 100 miles to 4 gallons per 100 miles, represents a 25 percent increase in fuel economy, from 20 mpg to 25 mpg.) The technology baseline for each vehicle class was set according to whether the majority of vehicles employed a given technology. Thus, all cars (but not trucks) are assumed to have four valves per cylinder and overhead camshafts even though a substantial number sold in the United States still have two valves, especially large cars.

The results of this technology assessment are summarized for passenger cars in Table 3-1, for SUVs and minivans in Table 3-2, and for pickup trucks in Table 3-3. The distinction between "production-intent" and "emerging" technologies for engines, transmissions, and vehicles is maintained.

For each technology considered, the tables give an estimated range for incremental reductions in fuel consumption (calculated in gallons per 100 miles). The ranges in fuel consumption improvement represent real-world variations that may result from many (sometimes competing) factors, including the baseline state of the engine, transmission, or vehicle; effectiveness in implementation; trade-offs associated with exhaust emissions, drivability, or corporate standards; trade-offs between price and performance; differences between new system design, on the one hand, and carryover or product improvement on the other; and other calibration or consumer acceptance attributes such as noise and vibration.

Similarly, the ranges of incremental cost in these tables represent variations to be expected depending on a number of conditions, including the difference between product improvement cycles and new component design; variations in fixed and variable costs, depending on manufacturer-specific conditions; commonality of components or subsystems across vehicle lines; and evolutionary cost reductions. In addition, since many of the cost figures were supplied by component and subsystem suppliers, a factor of 1.4 was applied to the supplied cost to arrive at the RPE to the consumer.

The analysis presented here is based on the average fuel consumption improvement and cost of each incremental technology, as shown in Tables 3-1, 3-2, and 3-3. For each vehicle class, the average fuel consumption improvement for the first technology selected is multiplied by the baseline fuel consumption (adjusted for the additional weight for safety improvements). This is then multiplied by the average improvement of the next technology, etc. Costs are simply added, starting at zero. Figures 3-4 to 3-13 show the incre-

TABLE 3-1 Fuel Consumption Technology Matrix—Passenger Cars

Baseline: overhead cams, 4-valve, fixed timing, roller finger follower.

	Fuel Consumption Improvement %	Retail Price Equivalent (RPE) ($)		Subcompact			Compact			Midsize			Large		
		Low	High	1	2	3	1	2	3	1	2	3	1	2	3
Production-intent engine technology																
Engine friction reduction	1-5	35	140	x	x	x	x	x	x	x	x	x	x	x	x
Low-friction lubricants	1	8	11	x	x	x	x	x	x	x	x	x	x	x	x
Multivalve, overhead camshaft (2-V vs. 4-V)	2-5	105	140												
Variable valve timing	2-3	35	140	x	x	x	x	x	x		x	x	x	x	x
Variable valve lift and timing	1-2	70	210		x	x		x	x		x	x		x	x
Cylinder deactivation	3-6	112	252											x	
Engine accessory improvement	1-2	84	112	x	x	x	x	x	x	x	x	x	x	x	x
Engine supercharging and downsizing	5-7	350	560									x			x
Production-intent transmission technology																
Five-speed automatic transmission	2-3	70	154	x			x			x	x		x	x	
Continuously variable transmission	4-8	140	350		x			x	x			x			x
Automatic transmission w/aggressive shift logic	1-3	—	70	x			x			x			x		
Six-speed automatic transmission	1-2	140	280								x			x	
Production-intent vehicle technology																
Aero drag reduction	1-2	—	140					x	x		x	x		x	x
Improved rolling resistance	1-1.5	14	56	x	x	x	x	x	x	x	x	x	x	x	x
Safety technology																
Safety weight increase	-3 to -4	0	0	x	x	x	x	x	x	x	x	x	x	x	x
Emerging engine technology																
Intake valve throttling	3-6	210	420		x			x			x			x	
Camless valve actuation	5-10	280	560			x			x			x			x
Variable compression ratio	2-6	210	490			x			x			x			x
Emerging transmission technology																
Automatic shift/manual transmission (AST/AMT)	3-5	70	280								x			x	
Advanced CVTs—allows high torque	0-2	350	840									x			x
Emerging vehicle technology																
42-V electrical systems	1-2	70	280						x		x	x		x	x
Integrated starter/generator (idle off-restart)	4-7	210	350			x			x			x			x
Electric power steering	1.5-2.5	105	150			x			x		x	x		x	x
Vehicle weight reduction (5%)	3-4	210	350			x			x			x			x

NOTE: An x means the technology is applicable to the particular vehicle. Safety weight added (EPA baseline + 3.5%) to initial average mileage/consumption values.

TABLE 3-2 Fuel Consumption Technology Matrix—SUVs and Minivans

Baseline (small SUV): overhead cams, 4-valve, fixed timing, roller finger follower.
Baseline (others): 2-valve, fixed timing, roller finger follower.

	Fuel Consumption Improvement (%)	Retail Price Equivalent (RPE) ($)		Small SUV			Mid SUV			Large SUV			Minivan		
		Low	High	1	2	3	1	2	3	1	2	3	1	2	3
Production-intent engine technology															
Engine friction reduction	1-5	35	140	x	x	x	x	x	x	x	x	x	x	x	x
Low-friction lubricants	1	8	11	x	x	x	x	x	x	x	x	x	x	x	x
Multivalve, overhead camshaft (2-V vs. 4-V)	2-5	105	140				x	x	x	x	x	x	x	x	x
Variable valve timing	2-3	35	140	x	x	x	x	x	x	x	x	x	x	x	x
Variable valve lift and timing	1-2	70	210		x	x		x	x		x	x		x	x
Cylinder deactivation	3-6	112	252				x	x	x	x	x			x	x
Engine accessory improvement	1-2	84	112	x	x	x	x	x	x	x	x	x	x	x	x
Engine supercharging and downsizing	5-7	350	560			x			x			x			x
Production-intent transmission technology															
Five-speed automatic transmission	2-3	70	154	x	x	x	x	x	x	x	x	x	x	x	x
Continuously variable transmission	4-8	140	350		x	x			x		x				
Automatic transmission w/aggressive shift logic	1-3	0	70	x			x			x	x		x		
Six-speed automatic transmission	1-2	140	280					x			x	x		x	x
Production-intent vehicle technology															
Aero drag reduction	1-2	0	140		x	x	x	x	x	x	x	x	x	x	x
Improved rolling resistance	1-1.5	14	56	x	x	x	x	x	x	x	x	x	x	x	x
Safety technology															
Safety weight increase	-3 to -4	0	0	x	x	x	x	x	x	x	x	x	x	x	x
Emerging engine technology															
Intake valve throttling	3-6	210	420		x			x			x			x	
Camless valve actuation	5-10	280	560			x			x			x			x
Variable compression ratio	2-6	210	490			x			x			x			x
Emerging transmission technology															
Automatic shift/manual transmission (AST/AMT)	3-5	70	280					x				x		x	
Advanced CVTs—allows higher torque	0-2	350	840						x						
Emerging vehicle technology															
42-V electrical systems	1-2	70	280		x	x		x	x		x	x		x	x
Integrated starter/generator (idle off–restart)	4-7	210	350		x	x		x	x		x	x		x	x
Electric power steering	1.5-2.5	105	150			x		x	x		x	x			
Vehicle weight reduction (5%)	3-4	210	350						x			x			x

NOTE: An *x* means the technology is applicable to the particular vehicle. Safety weight added (EPA baseline + 3.5%) to initial average mileage/consumption values.

TABLE 3-3 Fuel Consumption Technology Matrix—Pickup Trucks

Baseline: 2-valve, fixed timing, roller finger follower.

	Fuel Consumption Improvement (%)	Retail Price Equivalent (RPE) ($)		Small Pickup			Large Pickup		
		Low	High	1	2	3	1	2	3
Production-intent engine technology									
Engine friction reduction	1-5	35	140	x	x	x	x	x	x
Low-friction lubricants	1	8	11	x	x	x	x	x	x
Multivalve, overhead camshaft (2-V vs. 4-V)	2-5	105	140	x	x	x	x	x	x
Variable valve timing	2-3	35	140		x	x	x	x	x
Variable valve lift and timing	1-2	70	210		x	x		x	x
Cylinder deactivation	3-6	112	252	x	x			x	
Engine accessory improvement	1-2	84	112	x	x	x	x	x	x
Engine supercharging and downsizing	5-7	350	560			x		x	x
Production-intent transmission technology									
Five-speed automatic transmission	2-3	70	154	x	x		x	x	x
Continuously variable transmission	4-8	140	350			x			
Automatic transmission w/aggressive shift logic	1-3	0	70	x			x	x	
Six-speed automatic transmission	1-2	140	280		x			x	x
Production-intent vehicle technology									
Aero drag reduction	1-2	0	140	x	x		x	x	
Improved rolling resistance	1-1.5	14	56	x	x	x	x	x	x
Safety technology									
5% safety weight increase	–3 to –4	0	0	x	x	x	x	x	x
Emerging engine technology									
Intake valve throttling	3-6	210	420	x			x		
Camless valve actuation	5-10	280	560		x				x
Variable compression ratio	2-6	210	490		x				x
Emerging transmission technology									
Automatic shift/manual transmission (AST/AMT)	3-5	70	280	x					x
Advanced CVTs	0-2	350	840			x			
Emerging vehicle technology									
42-V electrical systems	1-2	70	280	x	x		x	x	
Integrated starter/generator (idle off–restart)	4-7	210	350	x	x		x	x	
Electric power steering	1.5-2.5	105	150	x	x		x	x	
Vehicle weight reduction (5%)	3-4	210	350						

NOTE: An *x* means the technology is applicable to the particular vehicle. Safety weight added (EPA baseline + 3.5%) to initial average mileage/consumption values.

mental results. Table 3-4 summarizes the results for the end point of each path. Further details of the methodology can be found in Appendix F.

No weight reduction was considered except in Path 3 for large passenger cars (Table 3-1) and medium and large SUVs (Table 3-2), where a 5 percent weight reduction is assumed as part of the 9- or 10-year forward projections.

The following example will assist the reader in understanding the relationship between Tables 3-1 through 3-3 and the results of the analysis, as summarized in Table 3-4. The average fuel consumption of midsize SUVs (based on 1999 sales-weighted averages) is 4.76 gal/100 miles, or 21.0 mpg. The assumed weight increase associated with future safety-enhancing features will result in a 3.5 percent increase in fuel consumption, to 4.93 gal/100 mile, or 20.3 mpg. This example is shown in the "Midsize SUV" column of Table 3-2 and the "Base FC w/Safety Weight" column of Table 3-4.

Within the Path 1 assumption for midsize SUVs, the technologies applied for incremental reductions in fuel consumption included engine friction reduction (1 to 5 percent); low-friction lubricants (1 percent); four-valve overhead cam (OHC) from two-valve overhead valve (OHV) or two-valve OHC (2 to 5 percent); variable intake valve timing (1 to 2 percent); cylinder deactivation (3 to 6 percent); engine accessory improvement (1 to 2 percent); five-speed automatic transmission (2 percent over four-speed automatic); aggressive shift logic (1 to 3 percent); and improved rolling resistance (1 to 1.5 percent). Multiplying the averages of these reductions yields a Path 1 estimated fuel consumption for midsize SUVs of 4.0 gal/100 miles, or 25.3 mpg. The incre-

TABLE 3-4 Estimated Fuel Consumption (FC), Fuel Economy (FE), and Incremental Costs of Product Development

Weight Class	Base FE	Base FC	Base FC w/Safety Weight	Average Fuel Economy (mpg)	Average Fuel Consumption (gal/100 mi)	Average Cumulative Cost ($)
Path 1						
Subcompact	31.3	3.19	3.31	34.7	2.88	465
Compact	30.1	3.32	3.44	33.4	2.99	465
Midsize	27.1	3.69	3.82	30.0	3.33	465
Large	24.8	4.03	4.17	27.9	3.58	675
Small SUV	24.1	4.15	4.29	26.7	3.74	465
Mid SUV	21.0	4.76	4.93	25.3	3.96	769
Large SUV	17.2	5.81	6.02	20.7	4.84	769
Minivan	23.0	4.35	4.50	26.5	3.78	587
Pickup—small	23.2	4.31	4.46	27.2	3.68	682
Pickup—large	18.5	5.41	5.59	21.2	4.71	587
Path 2						
Subcompact	31.3	3.19	3.31	37.5	2.67	1,018
Compact	30.1	3.32	3.44	36.6	2.73	1,088
Midsize	27.1	3.69	3.82	36.0	2.78	1,642
Large	24.8	4.03	4.17	34.5	2.90	2,167
Small SUV	24.1	4.15	4.29	31.4	3.18	1,543
Mid SUV	21.0	4.76	4.93	30.8	3.25	2,227
Large SUV	17.2	5.81	6.02	24.7	4.05	2,087
Minivan	23.0	4.35	4.50	34.0	2.94	2,227
Pickup—small	23.2	4.31	4.46	34.0	2.94	2,227
Pickup—large	18.5	5.41	5.59	28.2	3.55	2,542
Path 3						
SubCompact	31.3	3.19	3.31	43.9	2.28	2,055
Compact	30.1	3.32	3.44	42.9	2.33	2,125
Midsize	27.1	3.69	3.82	41.3	2.42	3,175
Large	24.8	4.03	4.17	39.2	2.55	3,455
Small SUV	24.1	4.15	4.29	36.5	2.74	2,580
Mid SUV	21.0	4.76	4.93	34.2	2.92	3,578
Large SUV	17.2	5.81	6.02	28.4	3.52	3,235
Minivan	23.0	4.35	4.50	36.6	2.73	2,955
Pickup—small	23.2	4.31	4.46	36.6	2.73	3,298
Pickup—large	18.5	5.41	5.59	29.5	3.39	2,955

mental costs are estimated to average $769. Table 3-4 shows the end points of all 3 paths for all 10 vehicle classes.

For the application of other incremental technologies in Path 2 and Path 3, only incremental benefits relative to prior technology applications—such as "variable valve lift and timing" added to "variable valve timing" alone—were assumed, in an attempt to eliminate double counting. Likewise, the incremental cost was added on top of the previous cost increment. Although not shown in the figures, the uncertainty increases with each step along the path. The level of uncertainty is quite high in Path 3.

For minivans, SUVs, and other light trucks, the potential for reductions in fuel requirements is much greater than for smaller passenger cars.

A review of the data shows that the technologies evaluated as emerging under Path 3 for the larger vehicles include camless valve actuation systems combined with a variable compression ratio and—possibly—clutched supercharging. A competing technology that could exceed the fuel consumption gains for such vehicles is turbocharged, intercooled, direct-injection diesel engines. This advantage is demonstrated by the growing popularity of such powertrains, which power almost all vehicles over 5,000 lb in Europe and 30 to 40 percent of all private passenger vehicles, even compact and subcompact classes.

However, the ability of these powertrains to meet federal for Tier 2 and California SULEV emissions standards is unproven, mainly because of shortcomings in exhaust gas aftertreatment. Therefore, the most efficient stoichiometric gasoline strategies (excluding hybrid vehicles, discussed below) were chosen to represent emerging technologies with proven thermodynamic advantages, which can employ relatively conventional three-way-catalyst aftertreatment systems.

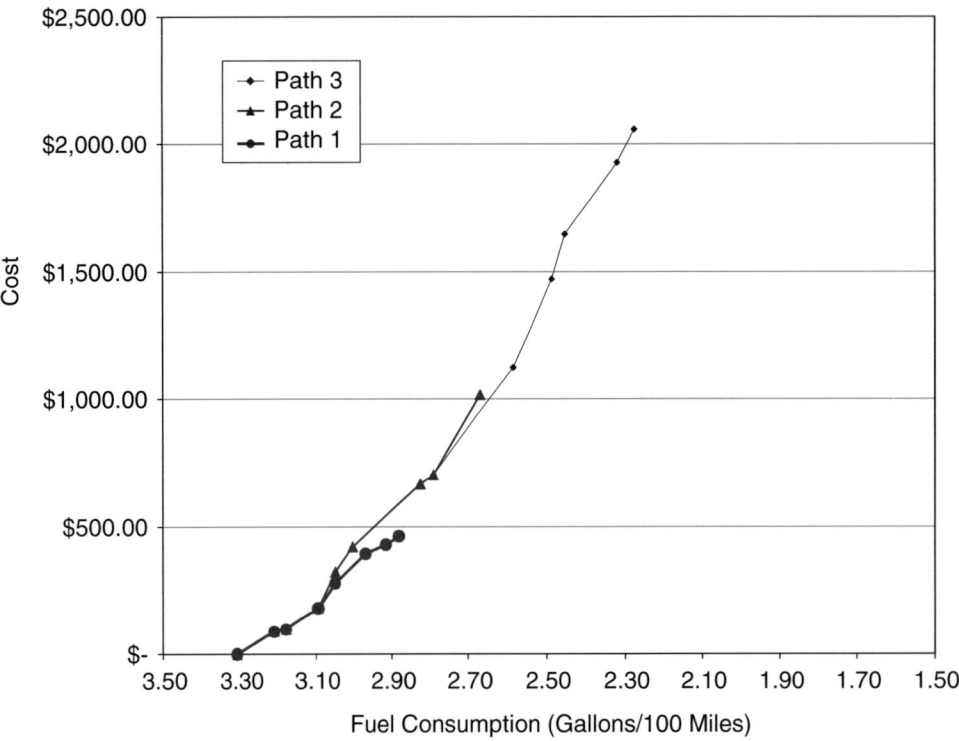

FIGURE 3-4 Subcompact cars. Incremental cost as a function of fuel consumption.

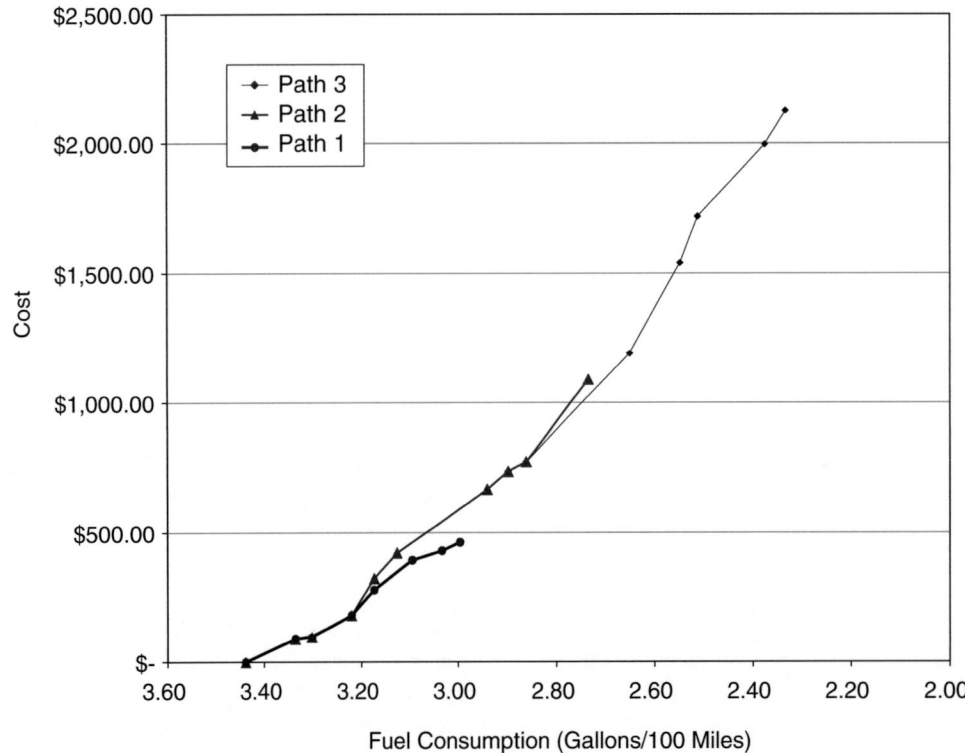

FIGURE 3-5 Compact cars. Incremental cost as a function of fuel consumption.

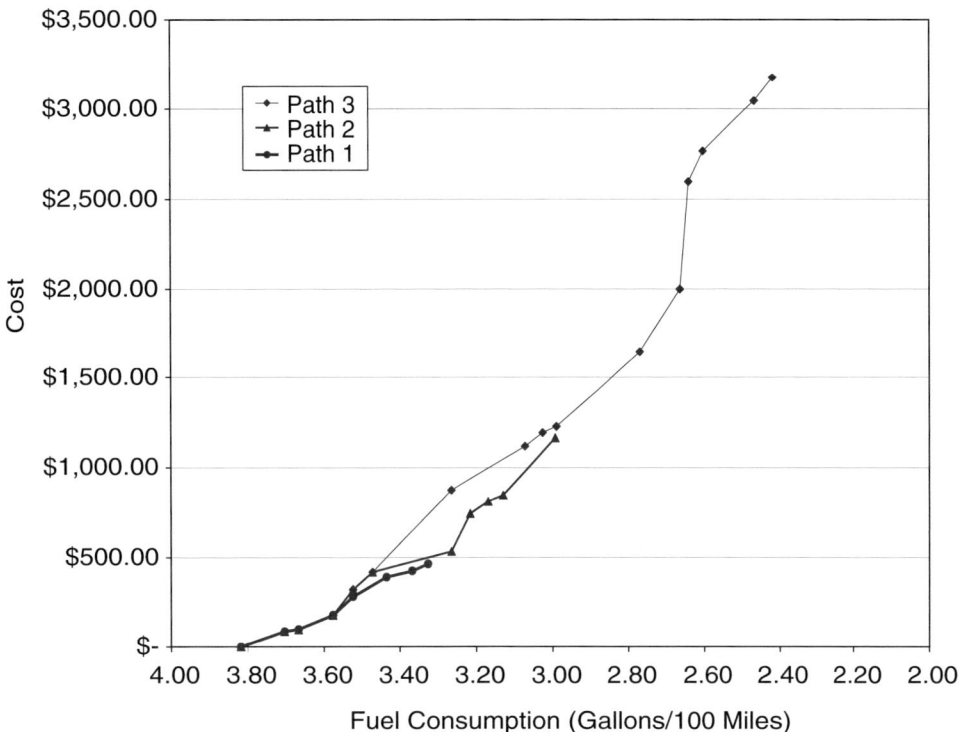

FIGURE 3-6 Midsize cars. Incremental cost as a function of fuel consumption.

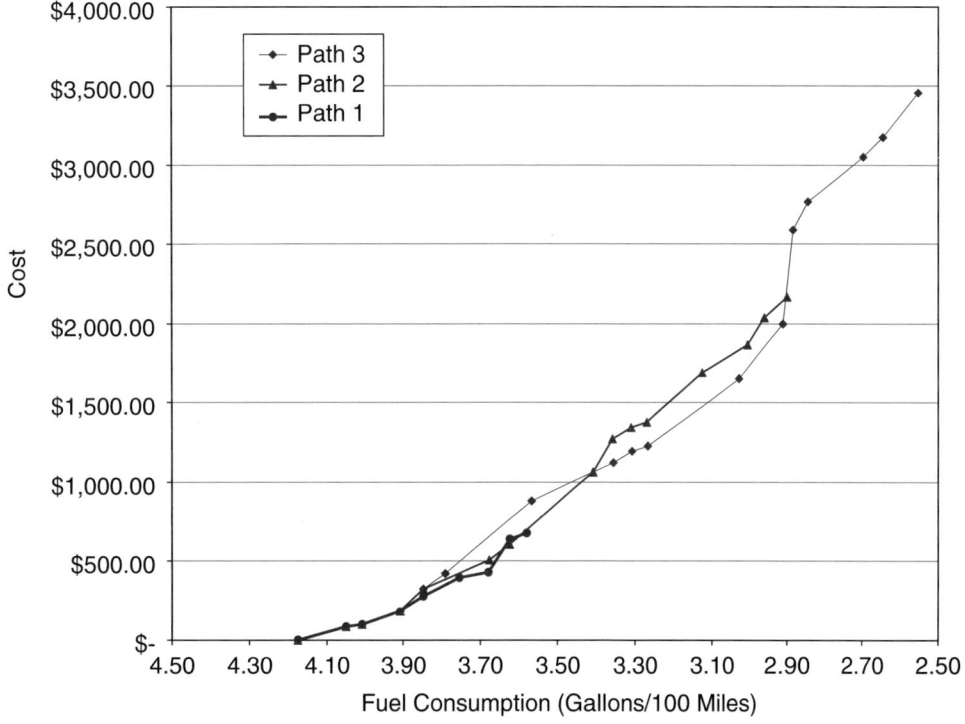

FIGURE 3-7 Large cars. Incremental cost as a function of fuel consumption.

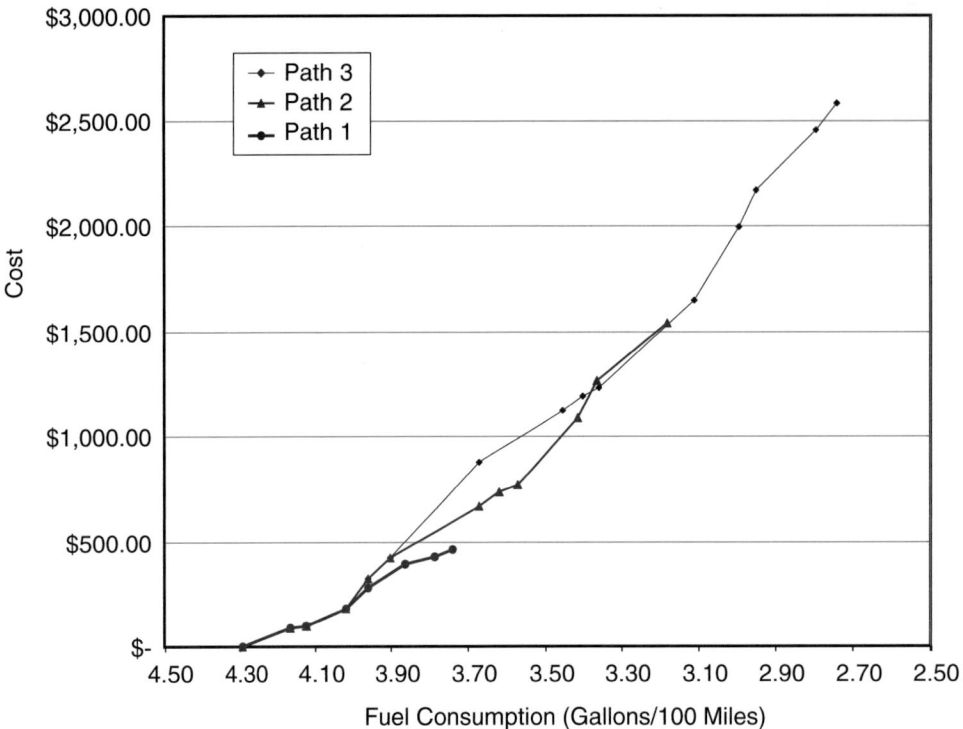

FIGURE 3-8 Small SUVs. Incremental cost as a function of fuel consumption.

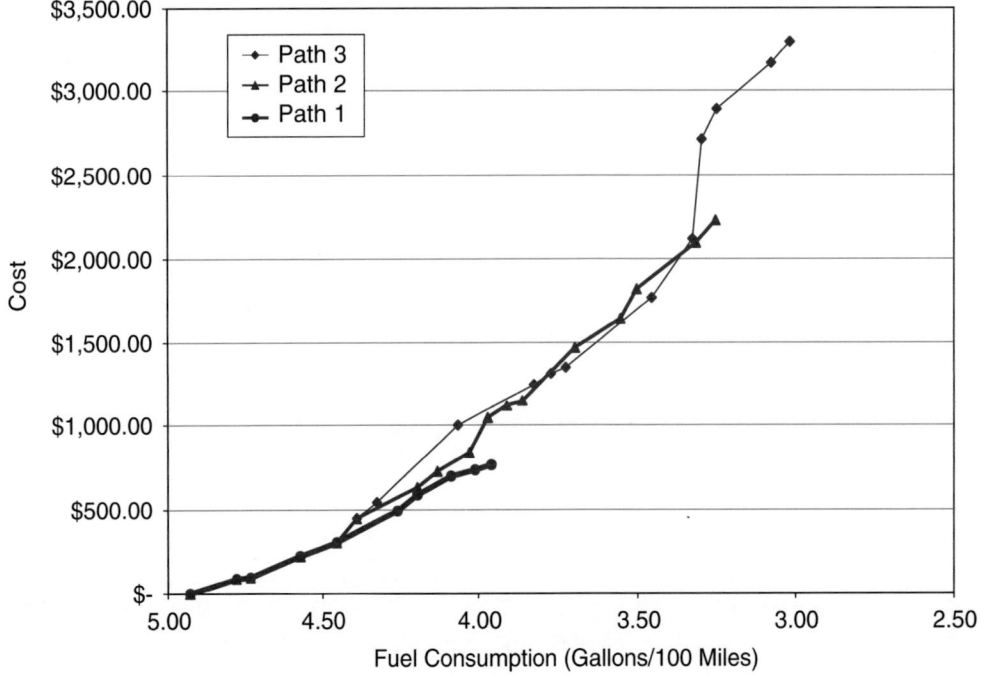

FIGURE 3-9 Midsize SUVs. Incremental cost as a function of fuel consumption.

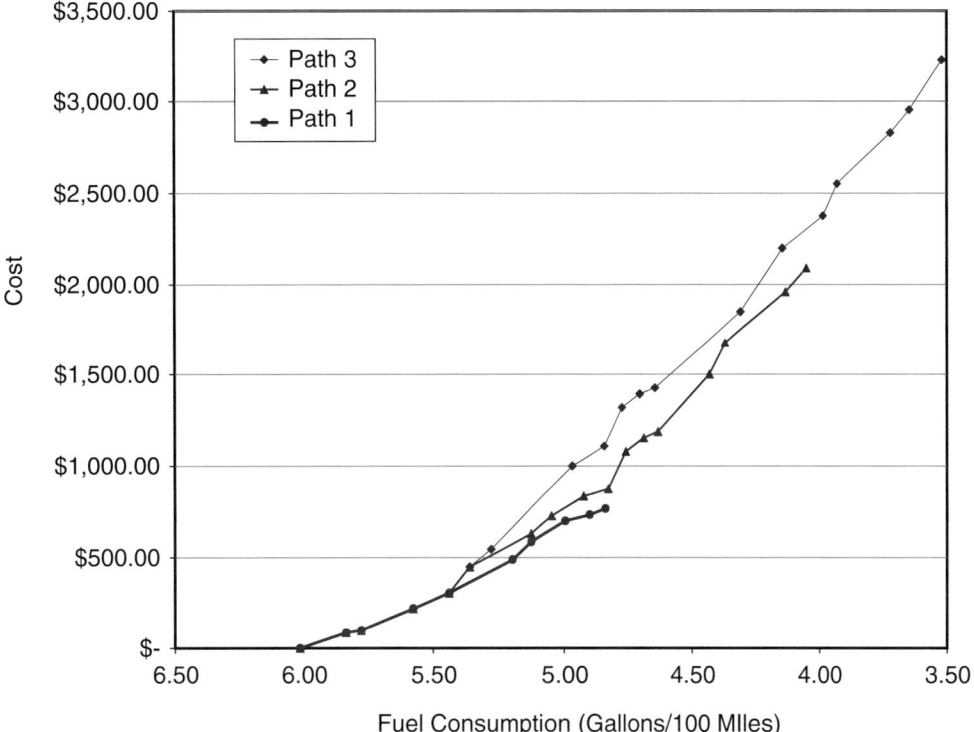

FIGURE 3-10 Large SUVs. Incremental cost as a function of fuel consumption.

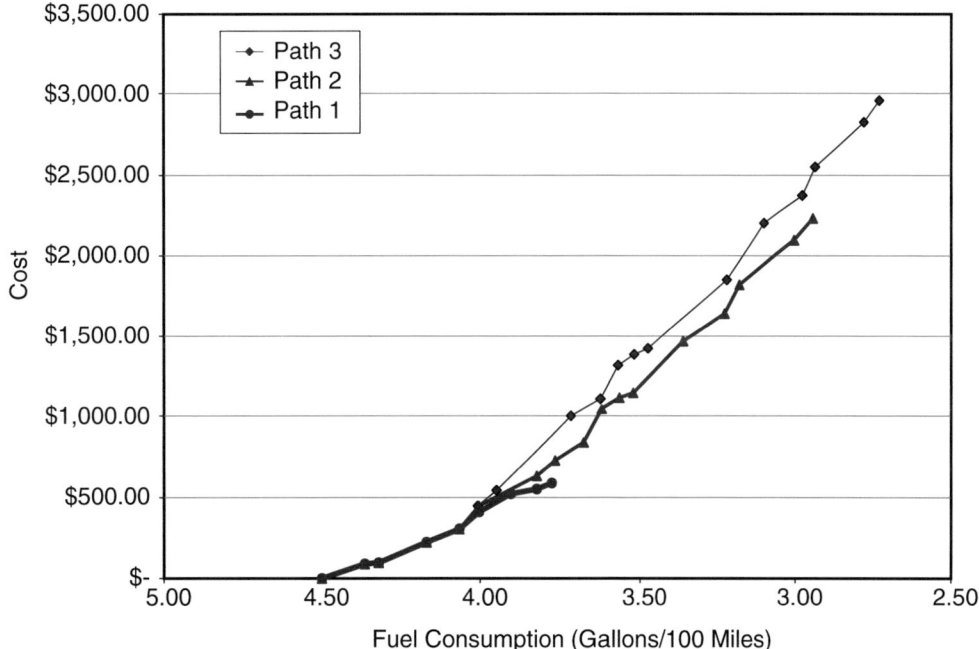

FIGURE 3-11 Minivans. Incremental cost as a function of fuel consumption.

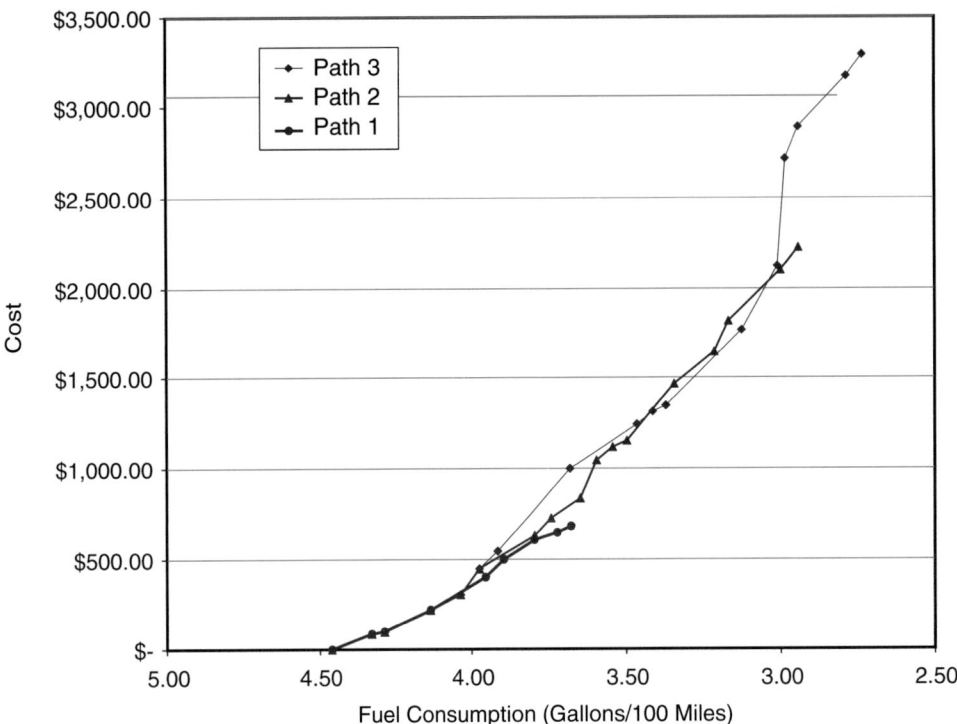

FIGURE 3-12 Small pickups. Incremental cost as a function of fuel consumption.

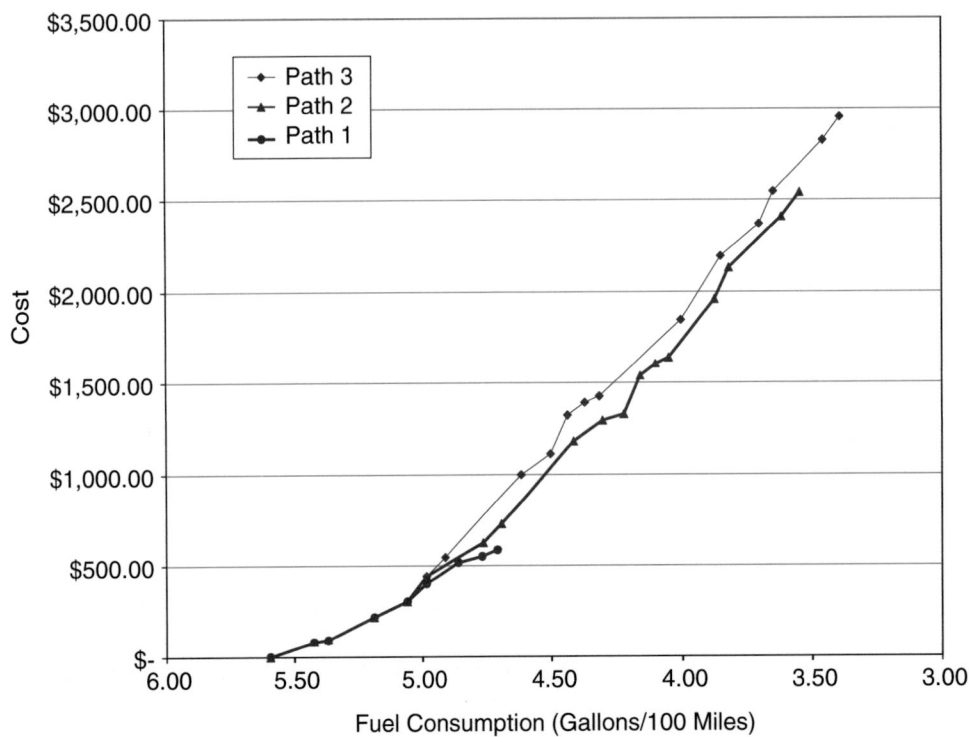

FIGURE 3-13 Large pickups. Incremental cost as a function of fuel consumption.

HYBRID VEHICLES

The concept of hybrid electric vehicles (HEVs) is not new, but the technology for application in vehicles advanced significantly in the last decade. Hybrid vehicles combine the power output of an internal combustion engine with a battery or other energy source to gain energy efficiency, reduce exhaust emissions, or, in some cases, improve acceleration. HEVs have received significant attention in the Partnership for a New Generation of Vehicles (PNGV). Ford, GM, and DaimlerChrysler have announced plans to introduce PNGV-developed hybrid technologies into production beginning in MY 2003. Toyota and Honda have already begun offering MY 2001 HEVs in limited production in the United States. Readers who desire further information on HEVs may consult a variety of sources, including *Automotive News* (2001) and An (2001), as well as a National Research Council report on the PNGV program (NRC, 2000a). Figure 3-14 shows the power relationship between an internal combustion engine and an electric motor for hybrid vehicles.

In the future, HEVs may be viable in situations involving increasingly stringent emissions regulations or demands to reduce oil dependence. For automotive use, however, two important parameters directly influence the practicality of the energy storage system: (1) the energy storage density (compared with liquid petroleum-derived fuel) and (2) the ability to rapidly receive or release stored energy.

Various energy storage techniques have been evaluated for light-duty vehicle applications, including batteries, capacitors, flywheels, and hydraulic/pneumatic systems. Despite restrictions in their energy transfer rate, batteries remain of high interest due to their relatively high power density.

A survey of key vehicle and powertrain characteristics is summarized in Table 3-5. Breakdowns of the relative contributions of powertrain modifications, including downsizing and dieselization, are shown in Figure 3-15. The data clearly show that vehicle and powertrain features can, on their own (that is, apart from true hybridization), result in significant fuel economy gains.

Ford has announced plans to introduce a hybrid version of its Escape SUV in MY 2003. It will reportedly have a four-cylinder engine and achieve about 40 mpg in the EPA combined (55/45) cycle. Its electric motor will reportedly produce acceleration performance similar to that of a larger six-cylinder engine (if the battery is sufficiently charged).

GM's hybrid powertrain, called ParadiGM, is reportedly being considered for application in several vehicles. It apparently combines a 3.6-liter V6 engine with a pair of electric motors. The engine shuts off at idle and low speed, and the vehicle can reportedly accelerate from a stop using only electric power.

DaimlerChrysler is apparently planning to offer a hybrid Dodge Durango SUV in 2003. It will reportedly combine a 3.9-liter V6 engine driving the rear wheels with an electric motor driving the front wheels. Its performance is reported to equal that of the current V8, four-wheel-drive version (again with sufficient battery charge) at a somewhat higher

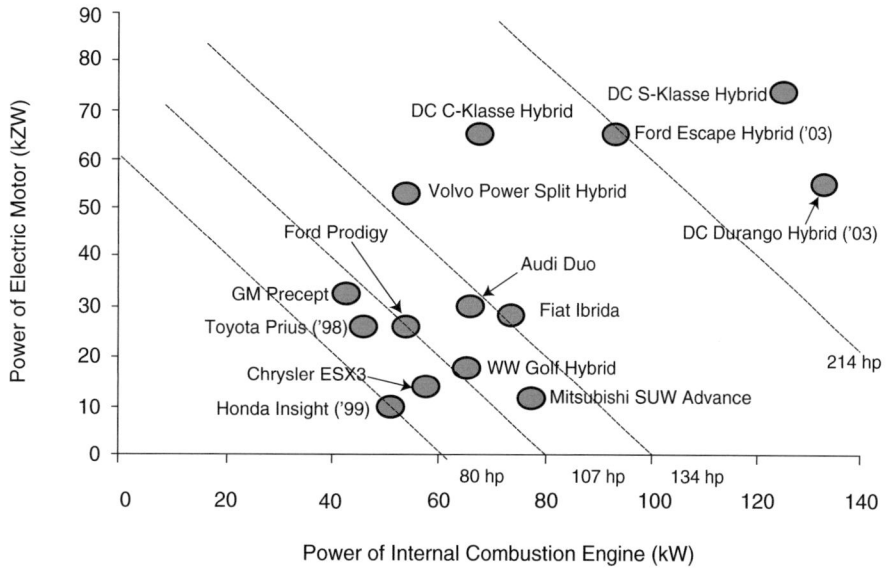

FIGURE 3-14 Relationship between the power of an internal combustion engine and the power of an electric motor in a hybrid electric vehicle.

TABLE 3-5 Published Data for Some Hybrid Vehicles

HEV Vehicle	Type	Status	Weight (lb)	Power Plant Type	Engine Size (L)	Engine Power (hp)	Battery Type	Motor Peak (kW)	Transmission Type	CAFE (mpg)[a]	0–60 (sec)	Data Source
Toyota Prius[b]	Gasoline hybrid	Com.	2,765	SI I-4	1.5	70	NiMH	33	CVT	58	12.1	[c]
Honda Insight	Gasoline hybrid	Com.	1,856	SI I-3	1.0	67	NiMH	10	M5	76	10.6	[d,e]
Ford Prodigy	Gasoline hybrid	Prot.	2,387	CIDI I-4	1.2	74	NiMH	16	A5	70	12.0	[d,f]
DC ESX3	Gasoline hybrid	Prot.	2,250	CIDI I-3	1.5	74	Li-ion	15	EMAT-6	72	11.0	[d,g]
GM Precept	Gasoline hybrid	Prot.	2,590	CIDI I-3	1.3	59	NiMH	35	A4	80	11.5	[d,h]

NOTE: SI, spark ignition; CIDI, compression ignition, direct injection; CVT, continuously variable transmission; M, manual; A, automatic; EMAT, electro-mechanical automatic transmission. SOURCE: An (2001).
[a] CAFE fuel economy represents combined 45/55 highway/city fuel economy and is based on an unadjusted figure.
[b] U.S. version.
[c] *EV News*, 2000, June, p. 8.
[d] NRC (2000a).
[e] *Automotive Engineering*, 1999, October, p. 55.
[f] The starter/generator rated 3 kW continuous, 8 kW for 3 minutes, and 35 kW for 3 seconds. We assume 16 kW for a 12-s 0–60 acceleration.
[g] *Automotive Engineering*, 2000, May, p. 32.
[h] Precept press release; the front motor is 25 kW and the rear motor is 10 kW, so the total motor peak is 35 kW.

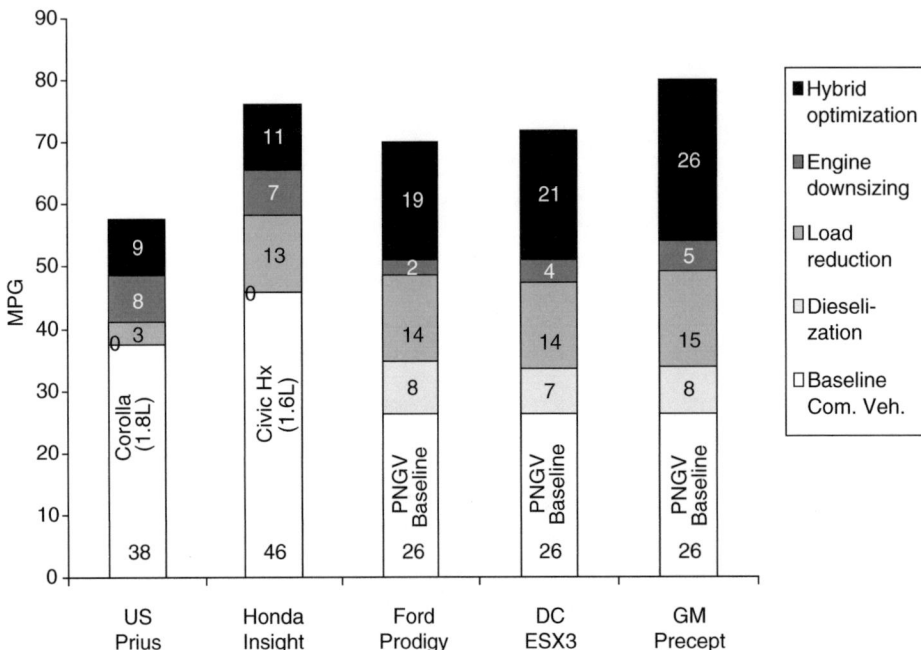

FIGURE 3-15 Breakdown of fuel economy improvements by technology combination. SOURCE: An (2001).

cost, but it will achieve fuel economy similar to its V6, two-wheel-drive equivalent.

Advanced HEVs cost much more than more conventional vehicles. In addition, overall system efficiencies must continue to improve, especially energy conversion, power transfer, electrochemical battery storage, and power output from the motors. These developments will allow greater overall fuel efficiency and system trade-offs that would result in reduced battery and motor sizes, extended electric-only propulsion range, improved power density, and reduced vehicle weight.

During the early introduction of these technologies, several obstacles must be addressed. First, warranty periods must be defined and, hopefully, extended with time. Second, the rate at which battery power systems can accept energy generated during a hard regenerative braking event must be improved. Finally, the potential safety consequences of a depleted battery (loss of acceleration power) must be clarified.

FUEL CELLS

The emerging technology of fuel cells is also receiving increasing attention and R&D funding on the basis of its potential use in passenger vehicles. A few concept vehicles are now in operation, and a few commercial vehicles may appear in niche markets in the next few years. Figure 3-16 schematically represents their principle of operation, using hydrogen as a fuel. Hydrogen enters the fuel cell through the porous anode. A platinum catalyst, applied to the anode, strips the electrons from the hydrogen, producing a positive hydrogen ion (a proton). The electrons pass through the load to the cathode as an electric current. The protons traverse the electrolyte and proceed to the porous cathode. Ultimately, through the application of a catalyst, the protons join with oxygen (from air) and the electrons from the power source to form water.

Different types of fuel cells employ different materials. According to Ashley (2001), "the proton exchange membrane (PEM) variety has emerged as the clear favorite for automotive use." Another type, the solid oxide fuel cell (SOFC), considered by Ashley and others as a less likely alternative, is represented by the alkaline air cell. The biggest difference between the SOFC and PEM technologies is their operating temperatures. While PEM cells run at 80°C, SOFC units run at 700° to 1000°C.

If hydrogen is used as the fuel, no atmospheric pollutants are produced during this portion of the energy cycle of fuel cells, since water is the only by-product. No hydrocarbons, carbon monoxide, NO_x, or particulates are produced. It is important to note that hydrogen could also be used as the fuel for an internal combustion engine. Louis (2001) quotes BMW as stating that a "spark ignition engine running on hydrogen is only slightly less efficient than a direct hydrogen fuel cell." The internal combustion engine will produce a certain level of NO_x during combustion. However, due to

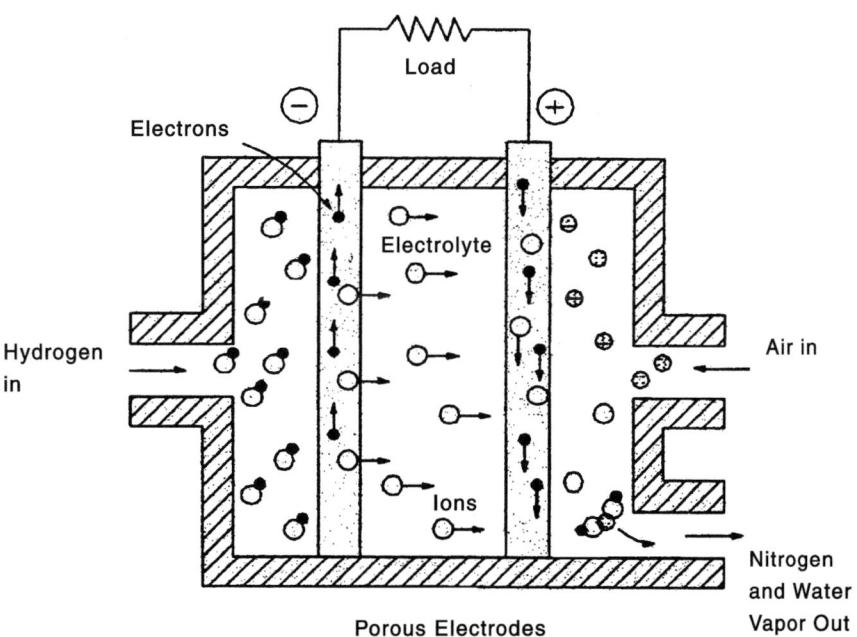

FIGURE 3-16 Working principles of a PEM fuel cell.

the very lean flammability limit of hydrogen, the NO_x concentration will be much lower than when using normal hydrocarbon fuels.

Energy is consumed, however, and exhaust emissions are likely to be generated in producing, transporting, and storing the hydrogen. The energy efficiency of a fuel cell cycle, "from well to wheels," includes energy losses and emissions from all of the steps of production, refining, and distribution of the fuel (see Attachment 4A). Since hydrogen is not naturally available, as are conventional fuels, it must be extracted from other hydrogen-containing compounds such as hydrocarbons or water. Unless the extraction is performed onboard, the hydrogen must be transported from the extraction point to the user. Hydrogen distribution systems are beyond the scope of this report, but this section evaluates two fuel cell systems that use a fuel reformer to generate hydrogen onboard the vehicle from either methanol (which can be produced from natural gas or biomass) or gasoline.

Onboard reformers have several common difficulties that must be overcome for commercial acceptance. They typically operate significantly above room temperature, with energy conversion efficiencies of 75 to 80 percent. The hydrogen is removed from the fuel by either catalysis or combustion. In addition, optimal operation occurs at process pressures above atmospheric. Furthermore, the response time and transient power requirements for vehicle application necessitate some form of onboard storage of hydrogen. For commercial success in passenger vehicles, the volume and weight parameters of the fuel reformer, fuel cell, and electric drive must be relatively competitive with current power trains and fuel tanks.

Using methanol as the liquid fuel offers the advantage of sulfur-free conversion (normal gasoline contains sulfur, which poisons fuel-cell stacks). Methanol has its own problems, however. It is toxic if ingested, highly corrosive, soluble in water (thereby posing a potential threat to underground water supplies), and currently relatively expensive to produce, compared with gasoline. However, proponents of methanol cite the toxicity of existing hydrocarbon fuels (gasoline or diesel fuel) and point out the low evaporative emissions, the absence of sulfur, and the potential for production from renewable sources.

All major automotive manufacturers are actively pursuing fuel cell systems using an onboard reformer. The committee believes, however, that advanced internal combustion engine-powered vehicles, including HEVs, will be overwhelmingly dominant in the vehicle market for the next 10 to 15 years. This conclusion is supported by the recently released study by Weiss et al. (2000).

Figure 3-17 shows the state of the art in fuel cell systems and the targets set by the Department of Energy for longer-term development. The figure shows that significant development advances are necessary to allow the fuel cell to become competitive with the internal combustion engine as a source of power. It is also important to note that the internal

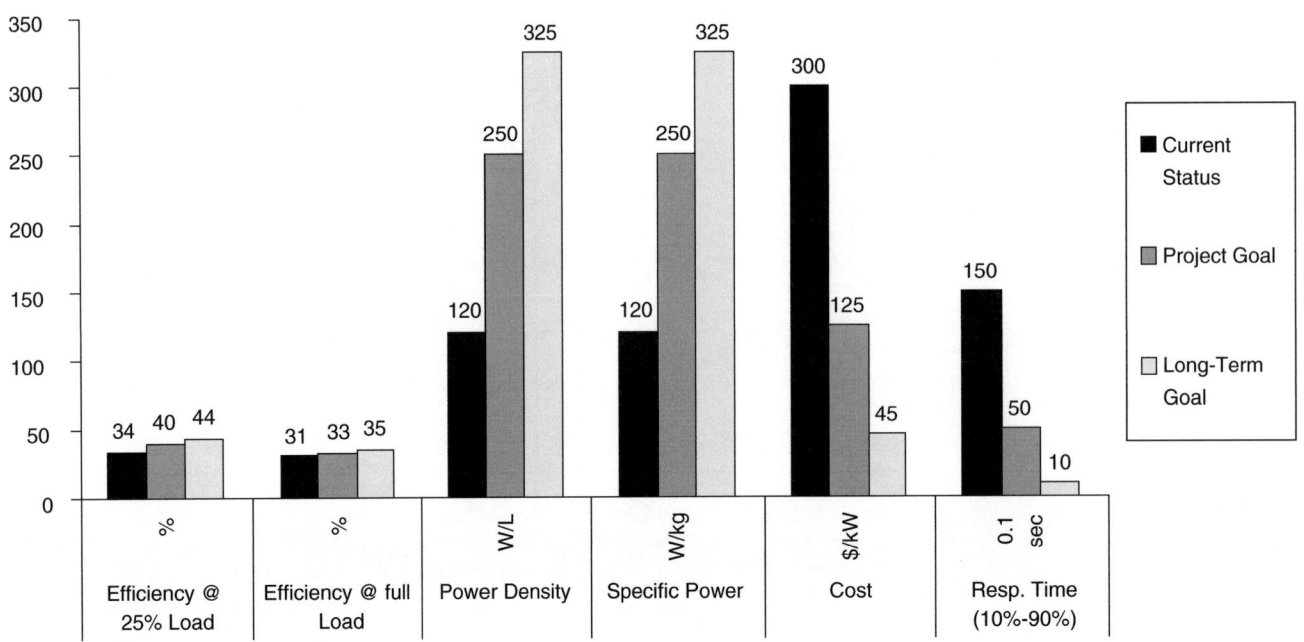

FIGURE 3-17 State of the art and future targets for fuel cell development.

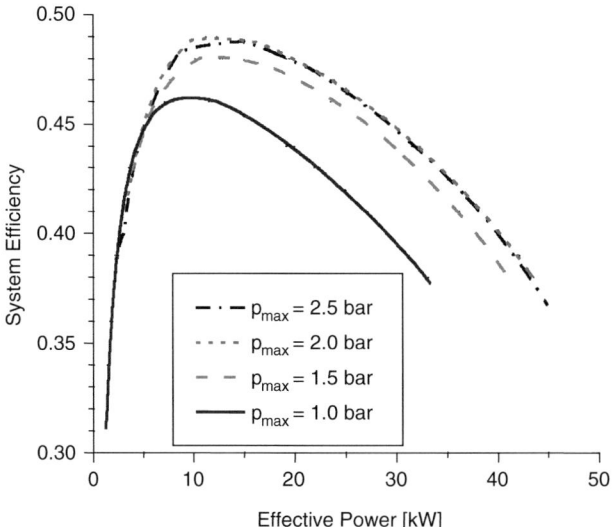

FIGURE 3-18 Typical fuel cell efficiency.

combustion engine will continue to advance as the fuel cell is being developed.

An additional issue that must be addressed is the reduced efficiency at higher loads, as shown in Figure 3-18. Substantial development will be required to overcome this characteristic and other challenges associated with power density, specific power, production cost, and system response time, before fuel cells can be successfully commercialized in an HEV.

REFERENCES

An, F. 2001. Evaluating Commercial and Prototype HEV's. SAE 2001-01-0951. Warrendale, Pa.: Society of Automotive Engineers (SAE).

Ashley, S. 2001. "Fuel Cells Start to Look Real." *Automotive Engineering International*. March.

CARB (California Air Resources Board). 1998. Particulate Emissions from Diesel-Fueled Engines As a Toxic Air Contaminant, November 3. Available online at <http://www.arb.ca.gov/toxics/dieseltac/dieseltac.com>.

CFR (Code of Federal Regulations). 2000. No. 40–Part 600, Office of the Code of Federal Regulations. National Archives and Records Administration, Washington, D.C., July.

Chon, D., and J. Heywood. 2000. Performance Scaling of Spark-Ignition Engines: Correlation and Historical Analysis of Production Engine Data. SAE Paper 2000-01-0565. Warrendale, Pa.: SAE.

DOE (Department of Energy). 2000. Fuel Cell and CIDI Engine R&D. Solicitation #DE-RP04-01AL67057, November.

Ecker, H.-J. 2000. Downsizing of Diesel Engines: 3-Cylinder/4-Cylinder. SAE Paper 2000-01-0990. Warrendale, Pa: SAE.

EEA (Energy and Environmental Analysis, Inc.). 2001. Technology and Cost of Future Fuel Economy Improvements for Light-Duty Vehicles. Final Report. Prepared for the committee and available in the National Academies' public access file for the committee's study.

EPA (Environmental Protection Agency). 1999. Regulatory Impact Analysis—Control of Air Pollution from New Motor Vehicles: Tier 2 Motor Vehicle Emissions Standards and Gasoline Sulfur Control Requirements. EPA 420-R99-023. Washington, D.C., December.

EPA. 2000a. Light-Duty Automotive Technology and Fuel Economy Trends 1975 Through 2000. EPA 420-R00-008. Washington, D.C., December.

EPA. 2000b. Regulatory Impact Analysis—Heavy-Duty Engine and Vehicle Standards and Highway Diesel Fuel Sulfur Control Requirements, December.

Leone, T.G., E.J. Christenson, and R.A. Stein. 1996. Comparison of Variable Camshaft Timing Strategies at Part Load. SAE Paper 960584. Warrendale, Pa.: SAE.

Louis, J. 2001. SAE 2001-01-01343. Warrendale, Pa.: SAE.

NRC (National Research Council). 1992. Automotive Fuel Economy: How Far Should We Go? Washington, D.C.: National Academy Press.

NRC. 2000a. Review of Research Program of the Partnership for a New Generation of Vehicles: Sixth Report. Washington, D.C.: National Academy Press.

NRC. 2000b. Review of the U.S. Department of Energy's Heavy Vehicle Technologies Program. Washington, D.C.: National Academy Press.

Pierik, R.J., and J.F. Burkhanrd. 2000. Design and Development of a Mechanical Variable Valve Actuation System. SAE Paper 2000-01-1221. Warrendale, Pa.: SAE.

Pischinger, M., W. Salber, and F. Vander Staay. 2000. Low Fuel Consumption and Low Emissions–Electromechanical Valve Train in Vehicle Operation. Delivered at the FISITA World Automotive Congress, June 12–15, Seoul.

Riley, R.Q. 1994. Alternative Cars in the 21st Century. Warrendale, Pa.: SAE.

Sierra Research. 2001. A Comparison of Cost and Fuel Economy Projections Made by EEA and Sierra Research. Prepared for the committee and available in the National Academies' public access file for the committee's study.

Weiss, M., J.B. Heywood, E.M. Drake, A. Schafter, and F.F. AuYeung. 2000. On the Road in 2020: A Life Cycle Analysis of New Automobile Technologies. Energy Laboratory, Massachusetts Institute of Technology, October. Cambridge, Mass.: MIT.

Wirbeleit, F.G., K. Binder, and D. Gwinner. 1990. Development of Pistons with Variable Compression Height for Increasing Efficiency and Specific Power Output of Combustion Engines. SAE Paper 900229. Warrendale, Pa.: SAE.

Zhao, Fu-Quan, and Ming-Chia Lai. 1997. A Review of Mixture Preparation and Combustion Control Strategies for Spark-ignited Direct-injection Gasoline Engines. SAE Paper 970627. Warrendale, Pa.: SAE.

Attachment 3A

A Technical Evaluation of Two Weight- and Engineering-Based Fuel-Efficiency Parameters for Cars and Light Trucks

Measuring the fuel economy of vehicles in miles per gallon (mpg) alone does not provide sufficient information to evaluate a vehicle's efficiency in performing its intended function. A better way to measure the energy efficiency of vehicles is needed, one that has a sound engineering basis. This attachment presents two weight-based parameters as examples of approaches that take the intended use of a vehicle into consideration. One is based on a vehicle's curb weight and the other includes its payload (passenger plus cargo). Because of the short time frame for the committee's study, an analysis sufficiently detailed to draw conclusions as to the value of these or other parameters was not possible.

MILES PER GALLON VERSUS GALLONS PER MILE AND HOW TO MEASURE

The physics of vehicle design can form the basis for parameters that more accurately represent system energy efficiencies and could be used by EPA in fuel economy testing. Mpg is not by itself a sufficient parameter to measure efficiency, since it is inherently higher for smaller vehicles and lower for larger vehicles, which can carry more passengers and a greater cargo load.

Although CAFE standards currently characterize vehicles by miles driven per gallon of fuel consumed, the inverse, gallons per mile, would be more advantageous for several reasons. As shown in Figure 3A-1, gallons per mile measures fuel consumption and thus relates directly to the goal of decreasing the gallons consumed. Note that the curve is relatively flat beyond 30 or 35 mpg because fuel savings become increasingly smaller as mpg increases. Also, the use of fuel consumption (gallons per mile) has analytical advantages, addressed in this attachment. To aid and clarify the analysis and make the numbers easier to comprehend, the term to be used is gallons per 100 miles (gal/100 miles). A vehicle getting 25 mpg uses 4 gal/100 miles.

For reproducibility reasons, fuel consumption measurements are made on a chassis dynamometer. The driving wheels are placed on the dynamometer rollers; other wheels do not rotate. Thus rolling resistance, as with aerodynamic drag, must be accounted for mathematically. Vehicle coast-down times are experimentally determined (a measure of aerodynamic drag); an auxiliary power unit (APU) ensures that dynamometer coast-down times are in reasonable agreement with road-tested coast-down times. Test reproducibility is in the few percent range. The driver follows two different specified cycles, city and highway, which were deduced from traffic measurements made some 30 to 40 years ago. A change in the test cycle is not a minor item—much engineering know-how is based on the present cycle, which is also used for exhaust emissions measurements.

WEIGHT-SPECIFIC FUEL CONSUMPTION

Figure 5A-4 (in Chapter 5) plots gal/100 miles versus vehicle weight for MY 2000 vehicles. The vertical scatter along a line of constant weight reflects the fact that vehicles

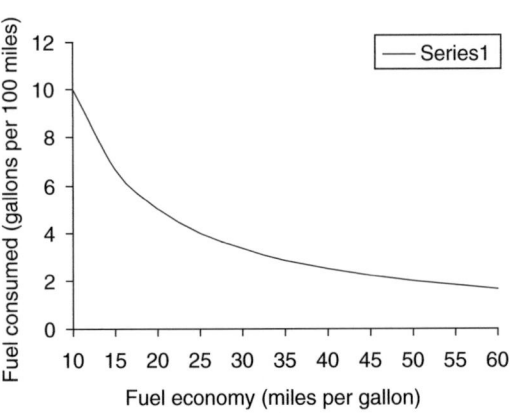

FIGURE 3A-1 Dependence of fuel consumption on fuel economy. SOURCE: NRC (2000).

of the same weight may differ in the efficiency of their drive trains or rolling resistance or aerodynamic drag (and thus in the number of gallons used to travel a given distance). While gal/100 miles is a straightforward parameter for measuring fuel consumption, it does not reflect the load-carrying capacity of the vehicle. Smaller cars, with lower fuel consumption, are designed to carry smaller loads, and larger cars and trucks, larger ones.

For engineering analysis purposes, it is convenient to normalize the data in Figure 5A-4, that is, divide the y value (vertical scale) of each data point by its curb weight in tons. The resulting new vertical scale is the weight-specific fuel consumption (WSFC). The units shown in Figure 3A-2 (and Figure 5A-5) are gal/ton of vehicle weight/100 miles. The straight horizontal line is a reasonable representation of the average efficiency of fuel use data for a wide variety of vehicle types and weights. It shows that the efficiency (WSFC) is approximately the same for this variety of different vehicle types (MY 2000, 33 trucks and 44 cars) and weights. Note that some vertical scatter is to be expected; all vehicles having approximately the same weight do not necessarily have the same drive-train efficiency.

Figure 3A-3 shows on-the-road data taken by *Consumer Reports* (April 2001). Their measurements were based on a realistic mixture of expressway, country-road, and city driv-

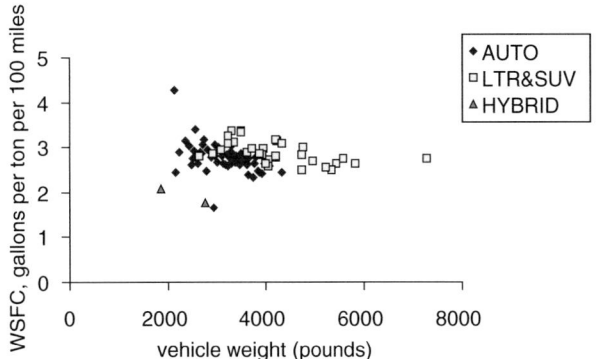

FIGURE 3A-3 Fleet fuel economy. Based on information from *Consumer Reports* (April 2001).

ing. Again, the efficiency of fuel use for their on-the-road tests is reasonably represented by a horizontal straight line. Figure 3A-3 illustrates the analytical utility of this approach. The lowest car point, at a little less than 3,000 lb, is a diesel engine; its WSFC is around 1.8 compared with around 2.7 for the average. The two hybrid points also show lower WSFC than the average but higher than the diesel.

Figure 3A-4 illustrates possible realistic reductions in WSFC. EPA has fuel-consumption data for more than a thou-

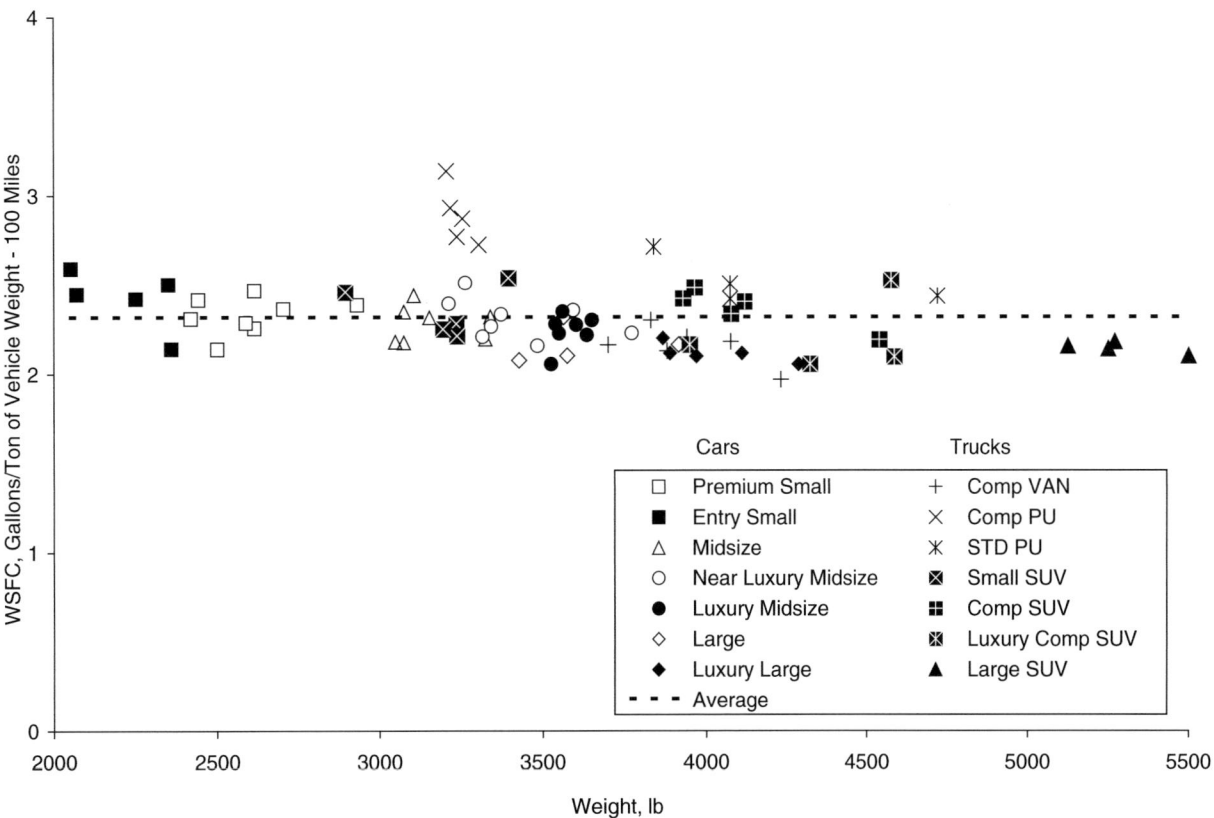

FIGURE 3A-2 Weight-specific fuel consumption versus weight for all vehicles.

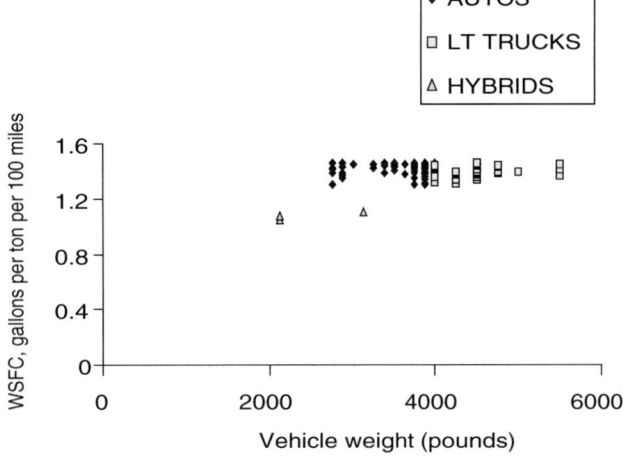

FIGURE 3A-4 Best-in-class fuel-efficiency analysis of 2000 and 2001 vehicles.

sand MY 2000 and 2001 vehicles. A horizontal line was drawn on a WSFC (highway) graph for these vehicles such that 125 vehicles were below this arbitrary line. The results for the 125 vehicles are shown in Figure 3A-4. The average WSFC value for all vehicles was 1.7; the average for the 125 vehicles was around 1.4. Since these were production vehicles, it would appear that application of in-production technologies to the entire fleet could produce significant reductions in WSFC.

LOAD-SPECIFIC FUEL CONSUMPTION

For a heavy-duty truck designed to carry a large payload, the most meaningful parameter would be normalized by dividing the gallons per mile by the tons of payload, to arrive at a load-specific fuel consumption (LSFC), that is, gallons per ton of payload per 100 miles. This number would be lowest for vehicles with the most efficient powertrain system and the least road load requirements (lightest weight, low accessory loads, low rolling resistance, and low drag) while moving the largest payload. Similarly, a reasonable parameter for a fuel-efficient bus would be gallons per passenger-mile.

The parameter LSFC is more difficult to define for light-duty vehicles than for heavy-duty trucks or buses, because the payloads are widely different (and harder to define) for these vehicles. This report calculates a total weight capacity by multiplying passenger capacity (determined by the number of seat belts) by an average weight per person (150 lb) and adding cargo weight capacity, which is the cargo volume multiplied by an average density (say, 15 lb/ft^3).[1] For pickup trucks, the difference between gross vehicle weight (GVW) and curb weight was used to determine payload. The weight of the passengers and cargo could be added to the vehicle's weight, and the sum used in the EPA fuel economy test to determine engine loading for the test cycle. Alternatively, the present fuel economy data could be used with the above average passenger plus cargo weight. The fuel consumption on the city and highway cycle would be measured and expressed as gallons/ton (passenger plus cargo weight)/100 miles.

Figure 3A-5 plots fuel economy against payload in tons for heavy-duty and light-duty vehicles. Lines of constant mpg-tons are also shown, with larger numbers representing more efficient transport of payloads. LSFC, the inverse of mpg-tons, is also shown on the lines, with lower numbers representing lower normalized vehicle fuel consumption. The point representing the PNGV goal is also shown.

Fuel economy measures based on this parameter would drive engineers to maximize the efficiency with which vehicles carry passengers and cargo while minimizing structural weight. This new fuel consumption parameter has the potential to be a better parameter to compare different types and sizes of vehicles.

Figure 3A-6 graphs the same light-duty vehicles' fuel consumption (in gal/100 miles) as a function of the payloads in Figure 3A-8. This figure shows the large difference in fuel consumption between cars (2.5 to 4 gal/100 mi) and trucks (3.5 to 5.5 gal/100 mi); the CAFE standards for both types of vehicles are included for reference. Figure 3A-7 shows LSFC (in gallons per ton of passengers plus cargo) for the same vehicles. LSFC appears to normalize fuel consumption, bridging all types of vehicles. Both types of vehicle—regardless of size and weight—are represented above and below the average line. This finding suggests that LSFC is a good engineering parameter for both cars and trucks.

COMPARING THE TWO WEIGHT-BASED PARAMETERS

The WSFC essentially normalizes the fuel consumed per 100 miles to take out the strong dependence on vehicle weight. Different weight vehicles can be compared more equitably. Lower WSFC parameters indicate lower road load requirements and/or higher powertrain efficiencies with lower accessory loads.

Figure 3A-9 shows fuel economy versus vehicle curb weight for the 87 light-duty vehicles. Constant efficiency lines (in mpg × tons of vehicle weight) are shown along with WFSC. Figure 3A-10 shows fuel economy versus vehicle payload for the 87 vehicles. The constant efficiency lines in mpg × tons of payload are shown along with LSFC. The utility of this plot is that it shows the interrelationship of fuel economy, payload, and LSFC in gallons/payload tons/100 miles. LSFC and WSFC show similar utility in determining whether certain types of vehicles are either above or below the average lines in Figures 3A-2 and

---

[1]This is an estimated density for cargo space. GM uses about 11 lb/ft^3 across a range of vehicles. Further study needs to be done to determine a representative design density to use.

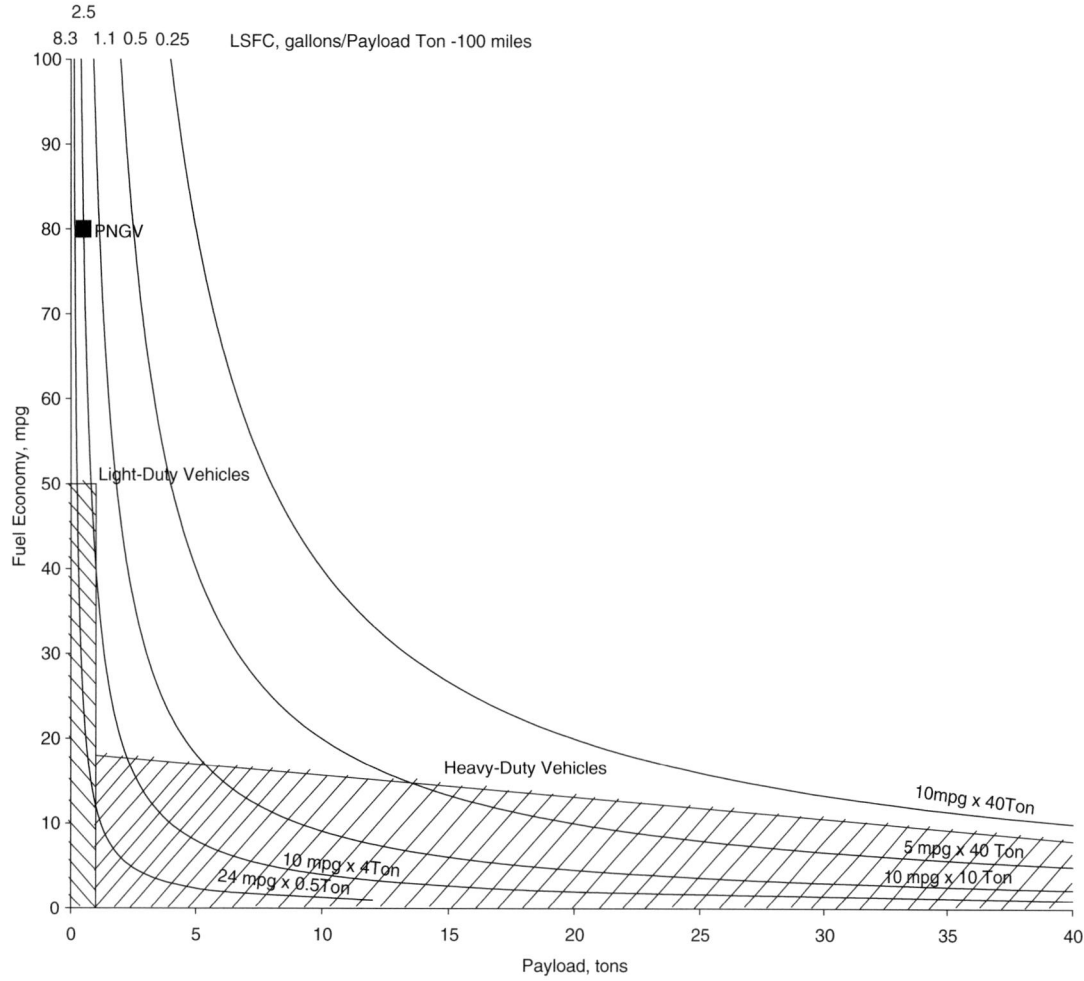

FIGURE 3A-5 LSFC versus payload for a variety of vehicles.

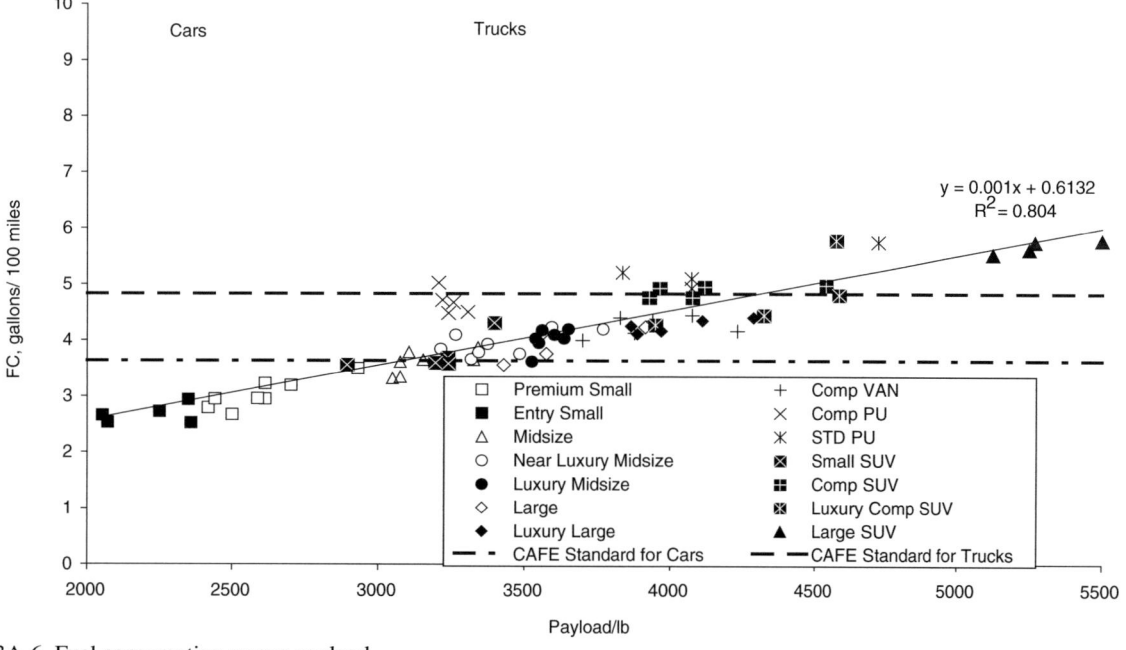

FIGURE 3A-6 Fuel consumption versus payload.

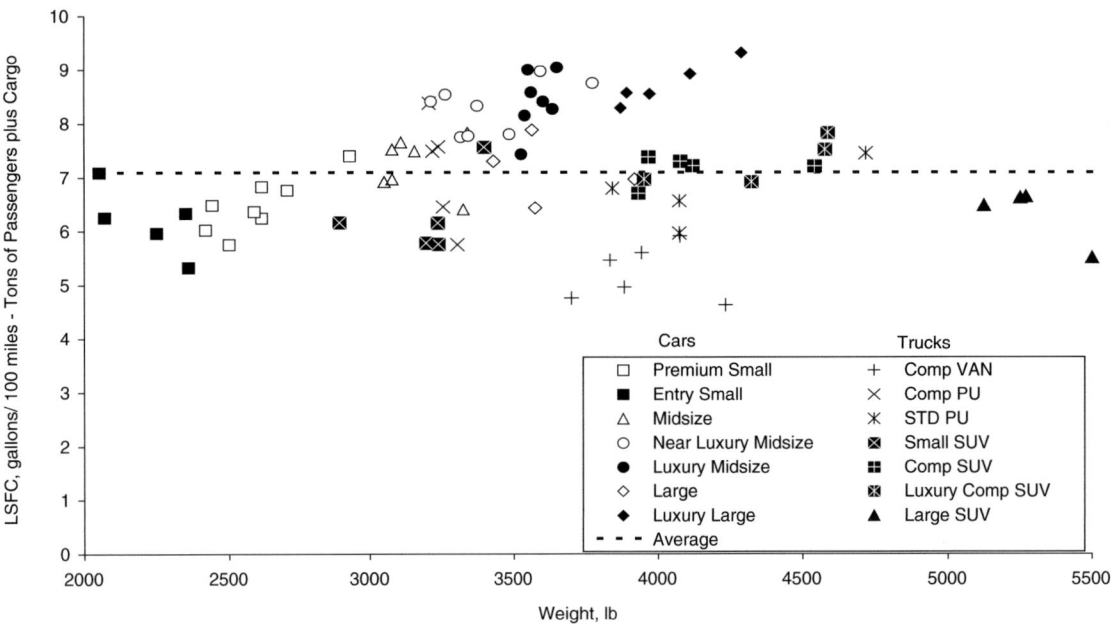

FIGURE 3A-7 Payload versus LSFC.

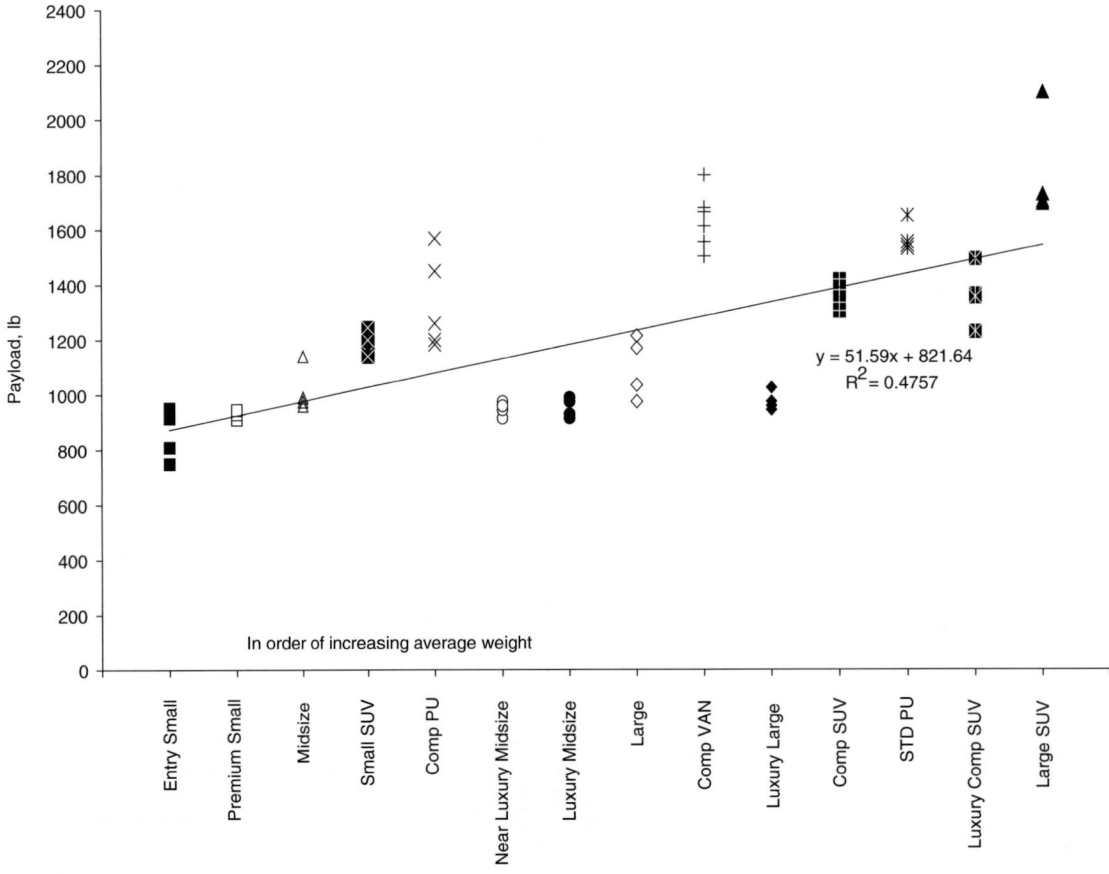

FIGURE 3A-8 Payload for a variety of vehicles.

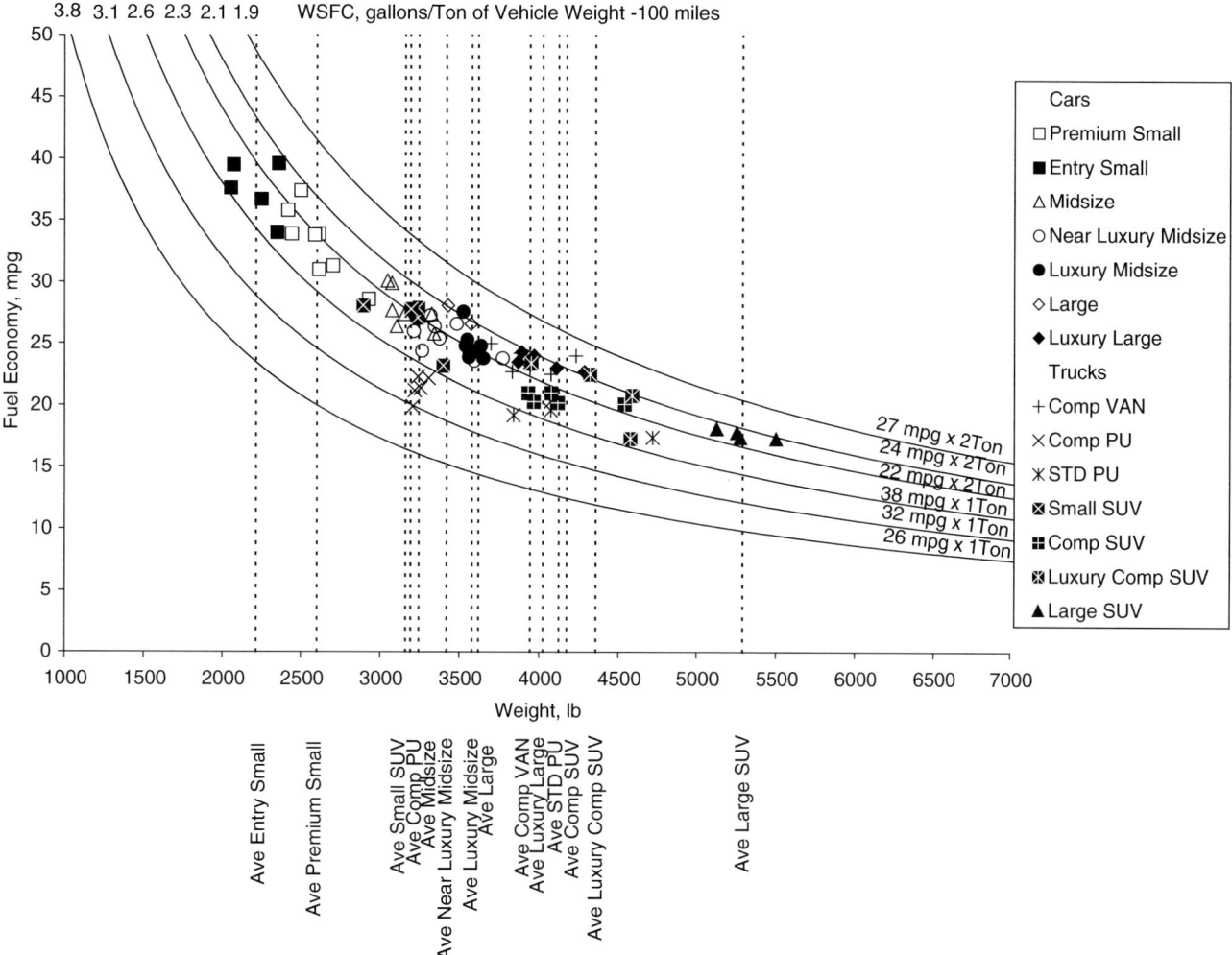

FIGURE 3A-9 Fuel economy as a function of average WSFC for different classes of vehicles.

3A-8. Using LSFC, however, would encourage manufacturers to consider all aspects of vehicle design, including materials, accessory power consumption, body design, and engine and transmission efficiency. The use of the LSFC number will show high-performance, heavy, two-seat sports cars without much cargo space and large luxury cars to be on the high side compared with vehicles designed to be fuel- and payload-efficient.

WSFC does not account for the load-carrying capacity of certain vehicles such as pickups, vans, and SUVs. The vans and large SUVs in Figure 3A-8 are shown below the average fit line and below the average WSFC line in Figure 3A-2, indicating they have highly efficient powertrain technologies and low road load/accessory load requirements. Pickups (PUs) are above the average WSFC line in Figure 3A-2, showing that it is difficult to design a truck that has low aerodynamic drag and is fuel efficient. When the fuel consumptions of these vehicles (vans, pickups, and SUVs) are normalized to pay-

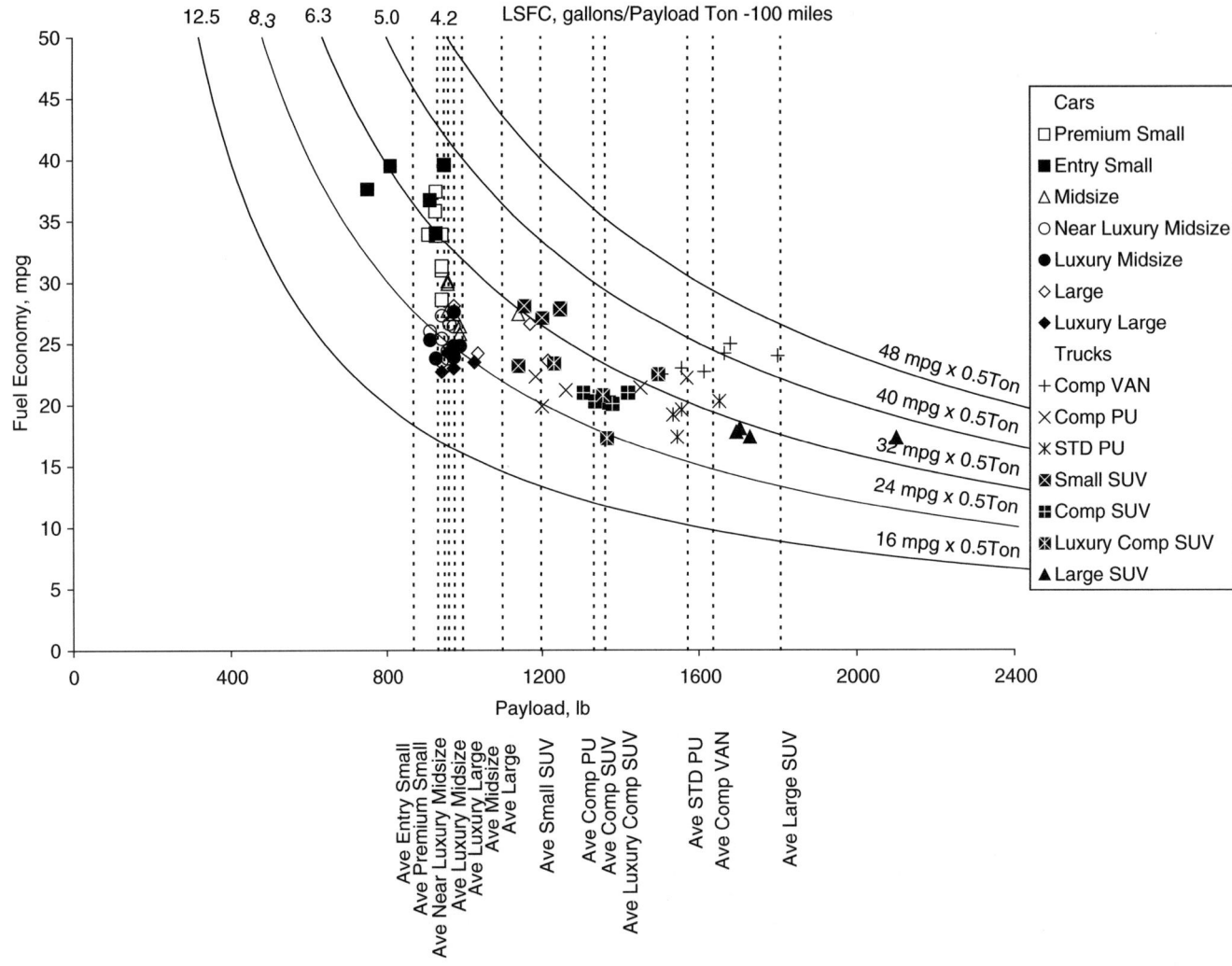

FIGURE 3A-10 Fuel economy versus average payload for different classes of vehicles.

load, they are below the average LSFC line in Figure 3A-8, indicating they are well designed for their intended use.

Conclusion

Both the WSFC and LSFC parameters have potential utility as fuel-efficiency parameters for vehicles, but their applicability requires additional study.

Reference

NRC (National Research Council). 2000. Automotive Fuel Economy: How Far Should We Go? Washington, D.C.: National Academy Press. p. 156.

4

Impact of a More Fuel-Efficient Fleet

If the technologies described in Chapter 3 are implemented, making vehicles more fuel efficient, there will be a variety of impacts. This chapter explores the potential impacts on energy demand and greenhouse gas emissions, cost-efficient levels of fuel economy, industry and employment, and safety.

ENERGY DEMAND AND GREENHOUSE GAS IMPACT

The fuel economy of light-duty cars and trucks and vehicle miles traveled (VMT) are the two most important factors underlying the use of energy and release of greenhouse gases in the light-duty fleet. Energy consumption during the manufacture of various components of the vehicles and their fuels is a lesser, but still significant, consideration in the life-cycle analyses of energy use and greenhouse gas emissions. Numerous projections were examined by the committee on possible future energy use and greenhouse gas emissions—for example, DeCicco and Gordon (1993), Austin et al. (1999), Charles River Associates (1995), EIA (2001), Patterson (1999), Greene and DeCicco (2000). To choose the best technology for overall energy efficiency, one must consider a "well-to-wheels" (WTW) analysis such as those that are described in Attachment 4A. The following discussion is designed to illustrate the impact of possible fuel economy changes and should *not* be interpreted in any way as a recommendation of the committee.

The committee calculated the potential magnitude of fuel savings and greenhouse gas emission reductions if new passenger car and light-truck fuel economy (in mpg) is increased by 15 percent, 25 percent, 35 percent, and 45 percent. These increases are assumed to be phased in gradually beginning in 2004 and to reach their full value in 2013. New vehicle sales shares of passenger cars and light trucks were held constant. Greenhouse gas emissions are for the complete WTW cycle based on the GREET model and include carbon dioxide, methane, and nitrogen oxides (Wang, 1996; Wang and Huang, 1999). The fuel for all light-duty vehicles is assumed to be gasoline, 70 percent conventional and 30 percent reformulated.

Allowance is made for a rebound effect—that is, a small increase in miles driven as fuel economy increases. Because of the long time required to turn over the fleet, the calculations were extended to the year 2030 to show the longer-range impacts of the increases in fuel economy. However, new vehicle mpg was held constant after 2013.

The base case approximates the 2001 Annual Energy Outlook forecast of the Energy Information Administration (EIA) except that the vehicle miles traveled are assumed to increase to 1.7 percent per year, slightly less than the 1.9 percent assumed by EIA. However, unlike the EIA forecast, fuel economy is assumed to remain constant. In the base case, annual gasoline use is projected to increase from 123 billion gallons (8 mmbd) in 2001 to 195 billion gallons (12.7 mmbd) by 2030 (Figure 4-1).

The magnitude of gasoline savings that could be achieved relative to the base case for various fuel economy increases between now and 2030 is shown in Figure 4-2. As a result of

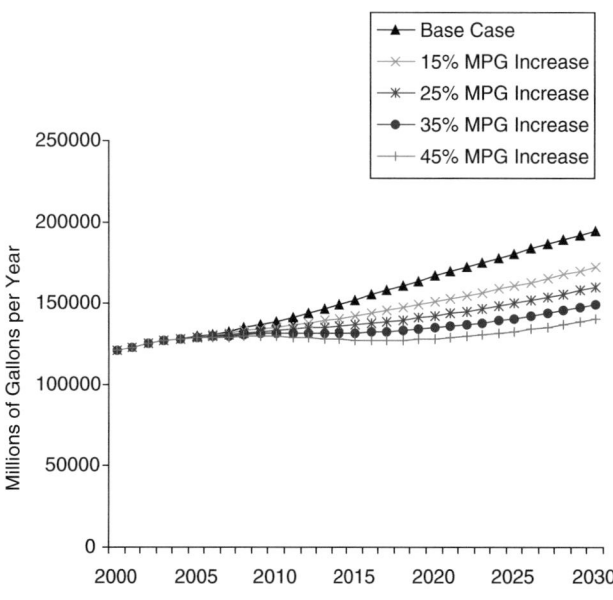

FIGURE 4-1 Fuel use in alternative 2013 fuel economy scenarios.

these hypothetical fuel economy increases, fuel savings in 2015 range from 10 billion gallons for a 15 percent fuel economy increase to 25 billion gallons for a 45 percent increase. By 2030 the fuel savings would be 22 billion gallons and 55 billion gallons for fuel economy increases of 15 percent and 45 percent, respectively. Since greenhouse gas emissions are correlated closely with gasoline consumption (Figure 4-1), they show a similar pattern (Figure 4-3). Growth in greenhouse gas emissions still occurs through 2030, but the growth is slower than it would have been without improved fuel economy (Figure 4-4).

ANALYSIS OF COST-EFFICIENT FUEL ECONOMY

The committee takes no position on what the appropriate level of fuel economy should be. The question, however, is often raised of how much investment in new technology to increase fuel economy would be economically efficient. That is, when does the incremental cost of new technology begin to exceed the marginal savings in fuel costs? Consumers might not choose to use this technology for fuel economy; they might choose instead to enhance other aspects of the vehicles. Such an estimate, however, provides an objective measure of how much fuel economy could be increased while still decreasing consumers' transportation costs. The committee calls this the cost-efficient level of fuel economy improvement, because it minimizes the sum of vehicle and fuel costs while holding other vehicle attributes constant.

The committee identified what it calls cost-efficient technology packages: combinations of existing and emerging technologies that would result in fuel economy improvements sufficient to cover the purchase price increases they would require, holding constant the size, weight, and performance characteristics of the vehicle(s).

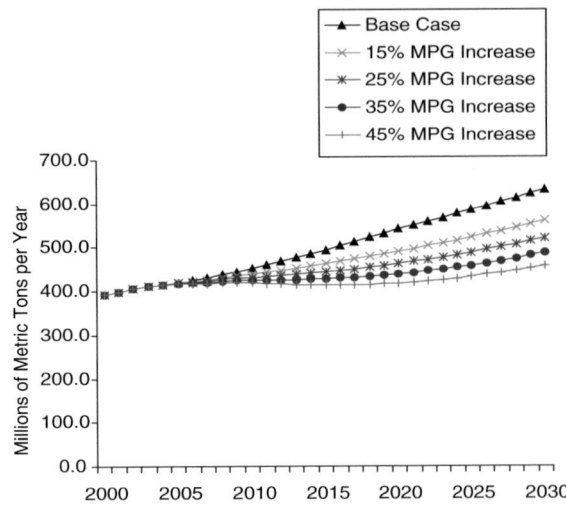

FIGURE 4-3 Fuel-cycle greenhouse gas emissions in alternative 2013 fuel economy cases.

The essence of analyzing cost-efficient fuel economy is determining at what fuel economy level the marginal costs of additional fuel-saving technologies equal the marginal benefits to the consumer in fuel savings. However, such analysis is conditional on a number of critical assumptions, about which there may be legitimate differences of opinion, including (1) the costs and fuel-efficiency effect of new technology and (2) various economic factors. The committee states its assumptions carefully and has investigated the effect of varying several key parameters.

Perhaps the most critical premise of this cost-efficient analysis is that key vehicle characteristics that affect fuel

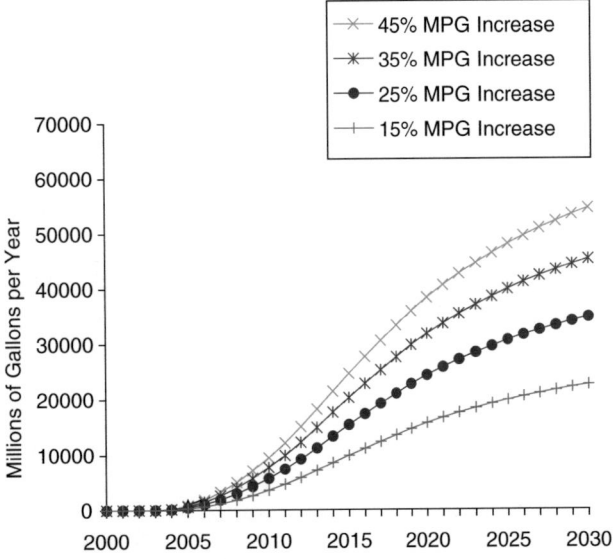

FIGURE 4-2 Fuel savings of alternative 2013 fuel economy improvement targets.

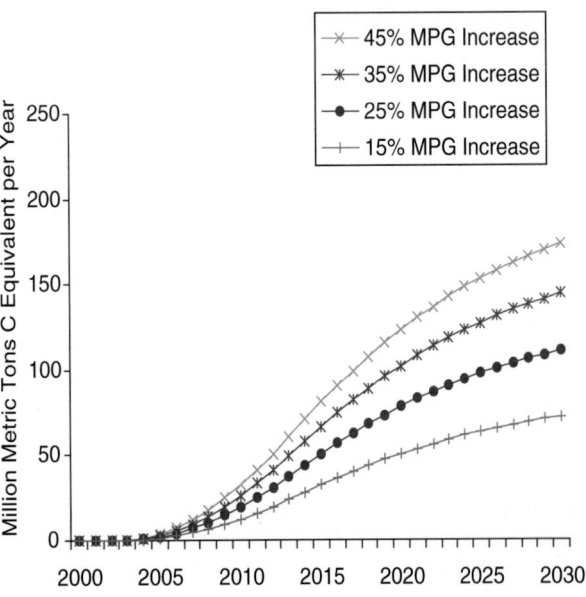

FIGURE 4-4 Greenhouse gas emissions reductions from hypothetical alternative fuel economy improvement targets.

economy are held constant: (1) acceleration performance, (2) size, in terms of functional capacity, (3) accessories and amenities, and (4) the mix of vehicle types, makes, and models sold. These factors are held constant for analytical clarity and convenience. Once one begins altering vehicle characteristics, the trade-off between fuel economy and cost becomes obscured in the myriad of other simultaneous changes of design and function that could be made. Since there is no obviously correct way to forecast future vehicle attributes and the sales mix, the committee holds them constant. In real-world markets, all of these factors will change over time and could have important implications for achieving a specific fuel economy goal.

Other key assumptions pertain to how consumers value fuel economy. The price for higher fuel economy technology is paid when a vehicle is purchased. Fuel savings accrue in the future, depending on how much a vehicle is driven and under what circumstances. Important uncertainties involve consumers' expectations about future fuel prices, the rates of return they will expect on investments in higher fuel economy, and their perception of how used-vehicle markets will value fuel economy. Not only are the average values of these key parameters uncertain, but they will certainly vary from consumer to consumer and from one market segment to another.

Incorporating all these uncertainties into the cost-efficient analysis would require a far more complex model and far more time to implement than the committee has available. Instead, the committee uses sensitivity analysis to illustrate the potential impacts of key assumptions on the outcomes of the cost-efficient analysis.

Despite the many uncertainties, the cost-efficient analysis is valuable for illustrating the range of feasible fuel economy improvement and the general nature of the cost of achieving higher fuel economy. However, it is critically important to keep in mind that the analysis is conditional on the assumptions of constant vehicle attributes and a number of key parameter values. Changing these assumptions would change the results, as the sensitivity analysis shows.

The cost-efficient analysis uses estimates of the costs of technologies and their impacts on fuel consumption in Chapter 3 (see Tables 3-1 through 3-3). Only technologies known to be capable of meeting future emissions standards and having either positive or small negative effects on other vehicle attributes were used. The technologies were reordered by cost-effectiveness, so that the order of implementation reflects increasing marginal cost per unit of fuel saved. Cost effectiveness is measured by the ratio of the midpoint fuel consumption reduction to the midpoint cost estimate. While this may not correspond to the actual order in which technologies are implemented by auto manufacturers, it is nonetheless the most appropriate assumption for analyzing the economic trade-off between vehicle price and fuel savings. The Path 3 technology scenario was used because all of the Path 3 technologies could be in full-scale production by 2013 to 2015. The data in Tables 3-1 through 3-3 were used to construct fuel economy supply functions and confidence bounds on those functions. The supply functions are combined with functions describing consumers' willingness to pay for higher fuel economy and are solved for the point at which the cost of increasing fuel economy by 1 mile per gallon equals its value to the consumer.

Three curves were constructed for each vehicle class to reflect the range of uncertainty shown in the tables in Chapter 3. The high cost/low fuel economy and low cost/high fuel economy curves provide a reasonable set of bounds for the average cost/average fuel economy curve. For the method used, see Greene (2001).

Consumers' willingness to pay is estimated using average data on vehicle usage, expected payback times, gasoline at $1.50 per gallon, and assumed rates of return on the consumer's investment in higher fuel economy. The key areas of uncertainty are the rate of return consumers will demand and the length of time over which they will value future fuel savings. Because each additional mile per gallon saves less fuel than the one before, the marginal willingness to pay for fuel economy will decline as fuel economy increases.

The cost-efficient point (at which the marginal value of fuel saved equals the marginal cost) is then found, assuming a fuel priced at $1.50 per gallon. This produces three fuel economy estimates for each vehicle class, reflecting the three curves. Finally, two cases are considered in which the key parameters of consumer discount rate and payback time are varied.

The following calculations for cost-efficient fuel efficiency are based on the key assumptions summarized in Table 4-1 for passenger cars and light-duty trucks. Two cases

TABLE 4-1 Key Assumptions of Cost-Efficient Analysis for New Car and Light Truck Fuel Economy Estimates Using Path 3 Technologies and Costs (see Chapter 3)

Assumption	Case 1	Case 2
First-year travel for new vehicles (mi/yr)	15,600	15,600
Rate of decrease in vehicle use (%/yr)	4.5	4.5
Payback time (yr)	14	3
Rate of return on investment (%)	12	0
Base fuel economy (mpg)		
Subcompact cars	31.3	31.3
Compact cars	30.1	30.1
Midsize cars	27.1	27.1
Large cars	24.8	24.8
Small SUVs	24.1	24.1
Midsize SUVs	21.0	21.0
Large SUVs	17.2	17.2
Minivans	23.0	23.0
Small pickups	23.2	23.2
Large pickups	18.5	18.5
On-road fuel economy (mpg) shortfall (%)	15	15
Effect of safety and emissions standards (%)	–3.5	–3.5

are developed. In Case 1 it is assumed that vehicles are driven 15,600 miles in the first year, decreasing thereafter at 4.5 percent/year, and that a gallon of gasoline costs $1.50 (1999 dollars). Payback time is the vehicle lifetime of 14 years. Base fuel economy is shown for each class of vehicles. It is assumed that there will be a 15 percent reduction in on-the-road gasoline mileage from the EPA combined test data. It is also assumed that there will be a 3.5 percent fuel economy penalty as a result of weight gains associated with future safety and emissions requirements and that consumers require a rate of return of 12 percent on the money spent on fuel economy. Because this last assumption is more subjective than the others, Case 2 was developed using a payback of 3 years at zero percent discount rate. This case represents the perspective of car buyers who do not value fuel savings over a long time horizon. Case 1 might be the perspective of policy makers who believe that the national interest is served by reducing fuel consumption no matter how many people own a vehicle over its lifetime.

The results of the cost-efficient analysis for Case 1 are shown in Table 4-2. Using the assumptions shown in Table 4-1, the calculation indicates that the cost-efficient increase in (average) fuel economy for automobiles could be increased by 12 percent for subcompacts and up to 27 percent for large passenger cars. For light-duty trucks, an increase of 25 to 42 percent (average) is calculated, with the larger increases for larger vehicles.

For example, Table 4-2 shows that a new midsize SUV typically (sales-weighted average) has a base fuel economy today of 21.0 mpg. The adjusted base (20.3 mpg) reflects the 3.5 percent fuel economy penalty for weight increases to meet future safety and emission standards. In the column labeled "Average," the cost-efficient fuel economy is 28.0 mpg. The percent improvement over the unadjusted base of 21.0 is 34 percent (shown in parentheses). The cost to obtain this improved fuel economy is estimated at $1,254.

However, there is wide uncertainty in the results. This is illustrated in the other two columns, which show an optimistic case (low cost/high fuel economy curve) and a pessimistic case (high cost/low fuel economy curve). In the low cost/high fuel economy column, the cost-efficient fuel economy increases to 30.2 mpg, for an improvement of 44 percent at a cost of $1,248. In the high cost/low mpg column, the fuel economy is 25.8 mpg, for an improvement of 23 percent at a cost of $1,589. In some cases the high cost/low mpg column will have a lower cost (and a significantly lower mpg) than the low cost/high mpg column because of the nature of the cost-efficient calculation and the relative slopes of the cost curves.

There is some evidence suggesting that consumers do not take a 14-year view of fuel economy when buying a new car. For that reason, Table 4-3 shows the cost-efficient fuel economy levels for 3-year payback periods. For cars, average cost-efficient levels are between 0.1 and 1.5 mpg higher than their respective adjusted base fuel economy levels, with the larger increases for the larger cars. The cost-efficient levels of the light-duty trucks are about 1.4 to 3.1 mpg higher than their respective adjusted base fuel economy levels, with the larger increases being associated with the larger trucks. The negative changes in fuel economy shown in Table 4-3 are because the base is used for this calculation. All vehicles still improve relative to the adjusted base, even in the high cost/low mpg column.

As shown in Table 4-2, for the 14-year payback (12 percent discount) case, the average cost-efficient fuel economy levels are between 3.8 and 6.6 mpg higher than their respective (unadjusted) bases for passenger cars, with the larger increases associated with larger cars. For light-duty trucks, the cost-efficient levels are about 6 to 7 mpg higher than the base fuel economy levels, with the larger increases associated with larger trucks.

The cost-efficient fuel economy levels identified in Tables 4-2 and 4-3 are *not* recommended fuel economy goals. Rather, they are reflections of technological possibilities and economic realities. Other analysts could make other assumptions about parameter values and consumer behavior. Given the choice, consumers might well spend the money required to purchase the cost-efficient technology packages on other vehicle amenities, such as greater acceleration, accessories, or towing capacity.

The fuel economy and cost data used in this study are compared with the data used in other recent studies in Figures 4-5 and 4-6 for cars and light trucks, respectively. The cost curves used in this study are labeled NRC 2001 Mid (average), NRC 2001 Upper (high cost/low mpg upper bound), and NRC 2001 Lower (low cost/high mpg lower bound). For comparison with two other studies, one by Sierra Research (Austin et al., 1999) and one by Energy and Environmental Analysis (EEA, 2001), the NRC curves were normalized to the sales-weighted average fuel economies of the new passenger car and light truck fleets. For passenger cars, the NRC average curve (NRC 2001 Mid) is similar to the Sierra curve (up to about 11 mpg increase) and slightly more optimistic than the EEA curve.

The Massachusetts Institute of Technology (MIT) (Weiss et al., 2000) and American Council for an Energy-Efficient Economy (ACEEE) (DeCicco et al., 2001) curves were obtained differently as they are based on complete, specific vehicles embodying new technology. The NRC, Sierra, and EEA analyses added technology incrementally. Neither the MIT nor the ACEEE studies present their results in the form of cost curves; the curves shown here are the committee's inferences based on data presented in those reports.

The MIT curve is calculated using the lowest cost and most fuel-efficient vehicles in the study, which uses advanced technology and a midsize sedan and projects to 2020. Similarly, the ACEEE-Advanced curves in Figures 4-5 and 4-6 are based on individual vehicles and advanced technology options. Both studies are substantially more optimistic than this committee's study, having used technology/cost options more advanced than those considered by the committee.

TABLE 4-2 Case 1: Cost-Efficient Fuel Economy (FE) Analysis for 14-Year Payback (12% Discount Rate)[a]

Vehicle Class	Base mpg[b]	Base Adjusted[c]	Low Cost/High mpg			Average			High Cost/Low mpg		
			FE mpg, (%)	Cost ($)	Savings ($)	FE mpg, (%)	Cost ($)	Savings ($)	FE mpg, (%)	Cost ($)	Savings ($)
Cars											
Subcompact	31.3	30.2	38.0 (21)	588	1,018	35.1 (12)	502	694	31.7 (1)	215	234
Compact	30.1	29.1	37.1 (23)	640	1,121	34.3 (14)	561	788	31.0 (3)	290	322
Midsize	27.1	26.2	35.4 (31)	854	1,499	32.6 (20)	791	1,140	29.5 (9)	554	651
Large	24.8	23.9	34.0 (37)	1,023	1,859	31.4 (27)	985	1,494	28.6 (15)	813	1,023
Light trucks											
Small SUVs	24.1	23.3	32.5 (35)	993	1,833	30.0 (25)	959	1,460	27.4 (14)	781	974
Mid SUVs	21.0	20.3	30.2 (44)	1,248	2,441	28.0 (34)	1,254	2,057	25.8 (23)	1,163	1,589
Large SUVs	17.2	16.6	25.7 (49)	1,578	3,198	24.5 (42)	1,629	2,910	23.2 (35)	1,643	2,589
Minivans	23.0	22.2	32.0 (39)	1,108	2,069	29.7 (29)	1,079	1,703	27.3 (19)	949	1,259
Small pickups	23.2	22.4	32.3 (39)	1,091	2,063	29.9 (29)	1,067	1,688	27.4 (18)	933	1,224
Large pickups	18.5	17.9	27.4 (48)	1,427	2,928	25.5 (38)	1,450	2,531	23.7 (28)	1,409	2,078

[a]Other key assumptions: See Table 4-1.
[b]Base is before downward adjustment of −3.5 percent for future safety and emissions standards.
[c]Base after adjustment for future safety and emissions standards (−3.5 percent).

TABLE 4-3 Case 2: Cost-Efficient Fuel Economy (FE) Analysis for 3-Year Payback (Undiscounted)[a]

Vehicle Class	Base mpg[b]	Base Adjusted[c]	Low Cost/High mpg			Average			High Cost/Low mpg		
			FE mpg, (%)	Cost ($)	Savings ($)	FE mpg, (%)	Cost ($)	Savings ($)	FE mpg, (%)	Cost ($)	Savings ($)
Cars											
Subcompact	31.3	30.2	33.3 (6)	180	237	30.3 (−3)	11	11	30.2 (−4)	0	0
Compact	30.1	29.1	32.3 (7)	202	268	29.1 (−2)	29	29	29.1 (−4)	0	0
Midsize	27.1	26.2	29.8 (10)	278	363	26.8 (−1)	72	76	26.2 (−4)	0	0
Large	24.8	23.9	28.2 (14)	363	488	25.4 (3)	173	190	23.9 (−4)	0	0
Light trucks											
Small SUVs	24.1	23.3	27.3 (13)	358	492	24.7 (2)	174	193	23.3 (−4)	0	0
Mid SUVs	21.0	20.3	25.0 (19)	497	721	22.7 (8)	341	407	20.3 (−4)	0	0
Large SUVs	17.2	16.6	21.1 (23)	660	992	19.7 (15)	567	740	18.3 (6)	373	424
Minivans	23.0	22.2	26.5 (15)	411	570	24.2 (5)	247	284	22.2 (−4)	0	0
Small pickups	23.2	22.4	26.9 (16)	412	579	24.4 (5)	247	285	22.4 (−4)	0	0
Large pickups	18.5	17.9	22.7 (23)	600	918	20.8 (12)	477	608	18.7 (1)	178	189

[a]Other key assumptions: See Table 4-1.
[b]Base is before downward adjustment of −3.5 percent for future safety and emissions standards.
[c]Base after adjustment for future safety and emissions standards (−3.5 percent).

In Figure 4-6, the committee's curve (NRC) is more optimistic than the Sierra curve and similar to the EEA curve. The ACEEE-Advanced curve, which is based on vehicles as discussed for cars above, is much more optimistic. It uses technology/cost options beyond those used for the committee's cost-efficient optimization.

POTENTIAL IMPACTS ON THE DOMESTIC AUTOMOBILE INDUSTRY

Regulations to increase the fuel economy of vehicles will require investments by automakers in R&D and tooling and will thus increase the costs of new vehicles. They will also

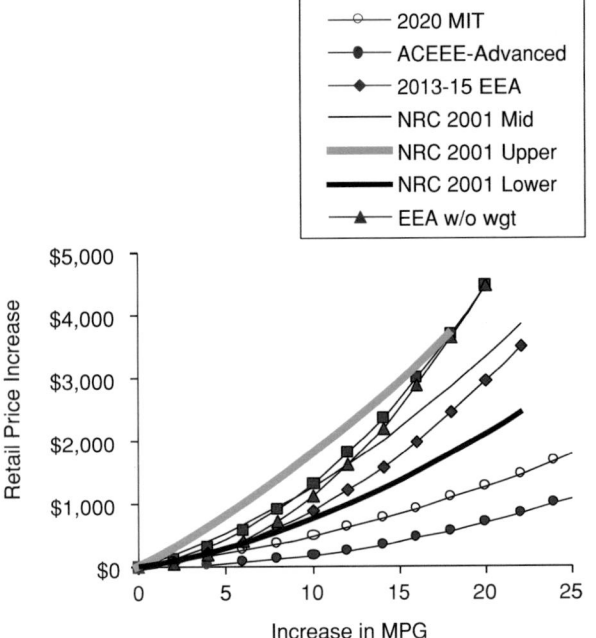

FIGURE 4-5 Passenger car fuel economy cost curves from selected studies.

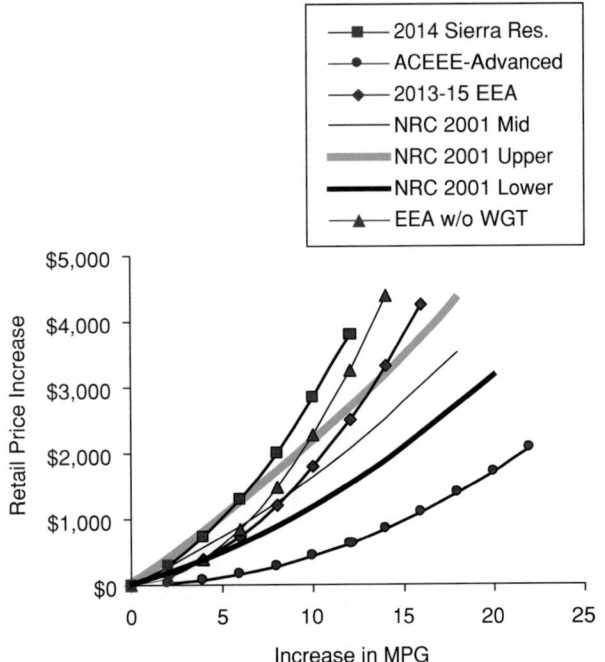

FIGURE 4-6 Light-truck fuel economy cost curves from selected studies.

divert resources that would otherwise go toward satisfying consumers' demands for performance, styling, and other vehicle features. No regulation is without cost to consumers or manufacturers. The impacts of CAFE standards in the past are reviewed in Chapter 2.

Looking Forward at the Automobile Market

As noted in Chapter 2, GM, Ford, and the Chrysler division of DaimlerChrysler will post lower profits in 2001 owing to the slower economy and the sharp rise in buyer incentives that the companies need to offer to maintain their market shares. Vehicle demand is expected to decline in 2001 from 17.4 million in 2000, with the final number depending in large part on the level of rebates and other incentives.

Foreign companies' share of the U.S. market has grown steadily and is now about 36 percent, compared with 26 percent in 1993. In 2000, GM, Ford, and Chrysler combined lost 3 percentage points in market share, and so far in 2001 have lost another 1.4 percentage points (despite per-unit incentives that are the highest in history and often triple the marketing support on competing foreign models). The gain in market share of foreign companies has accelerated in recent years as foreign manufacturers entered the light truck market with models that competed with traditional American pickups, minivans, and SUVs. They also created a new category, the crossover vehicle, which is built on a car chassis but looks like an SUV and may be classed as a light truck for fuel economy regulation. (Examples include the Toyota RAV-4 and Honda CRV.)

The erosion in profit margins and income at GM, Ford, and Chrysler began in 2000 as pricing pressures, once confined to passenger cars, spread into the light-truck sector and sales slowed. The supply of light trucks, until recently balanced against sharply rising demand, had allowed vehicle manufacturers to aggressively price these products. Although the Big Three still dominate the light-truck market with a 77 percent share, that is down from 86 percent in 1993 as a result of new products and capacity from Asian and European manufacturers. Additional North American truck capacity from Honda in Alabama and Toyota in Indiana and expansion of capacity by BMW and Nissan will add at least 750,000 units of new truck supply to the market over the next 3 years. Given the recent success of foreign manufacturers with new models and advanced technology (see Chapter 3), their share of the truck segment is likely to rise further in the coming years.

To cope with falling profits and lower cash balances, auto companies have cut back on discretionary spending, eliminated jobs through voluntary retirement plans, and now appear to be delaying product launch schedules. Standard & Poor's, the arbiter of creditworthiness, recently lowered its outlook on GM and Ford from stable to negative (Butters, 2001).

Even if vehicle demand rebounds in 2002 to the 17 million-unit level, industry profits are not expected to recover soon to historic levels. Increased competition could reduce truck profits to half of their former levels. If the proliferation of foreign brand crossover models draws buyers away from larger, heavier, and more expensive domestic models, the financial impact on the industry will be greater.

This is a difficult environment for GM, Ford, and the Chrysler division of DaimlerChrysler. In less than 3 years, Chrysler went from being the most profitable vehicle manufacturer in the country to a merger with Daimler that was forced by financial distress. While GM and Ford are in reasonable financial health, they cannot count on truck profits to generate above-average returns as they did in the past.

Nevertheless, the industry could adjust to possible changes in CAFE standards if they were undertaken over a long period of time, consistent with normal product life cycles. An abrupt increase in fuel economy standards (especially one that hurt the industry's ability to sell light trucks) would be more costly. A single standard that did not differentiate between cars and trucks would be particularly difficult to accommodate.

Criteria for Judging Regulatory Changes

The impacts on industry of changing fuel economy standards would depend on how they were applied. Some of the more important criteria are discussed here.

Timing and Scale of Increase

Raising standards too steeply over too short a time would require manufacturers to absorb much of the cost of obtaining or developing technology and tooling up to produce more efficient vehicles. It would lead to sharp increases in costs to manufacturers (and thus consumers). The benefits of precipitous increases in CAFE standards would flow instead to machine tool companies, component manufacturers, and developers of vehicle technology. Automakers would be forced to divert funds and talent away from longer-term investments (such as the PNGV). Given sufficient planning time, however, industry can adapt. Chapter 3 discusses the costs and timing involved in introducing fuel-saving technology. Generally, little change can be expected over the next few years, and major changes would require a decade.

Equivalence of Impact

If new regulations favor one class of manufacturer over another, they will distribute the costs unevenly and could evoke unintended responses. In general, new regulations should distribute the burden equally among manufacturers unless there is a good reason not to. For example, raising the standard for light trucks to that of cars would be more costly for light-truck manufacturers. On the other hand, tightening the standards for passenger vehicles while leaving light trucks alone would favor another set of manufacturers. A current proposal to simply increase the standard for light-duty trucks to the current level of the standard for passenger cars would operate in this inequitable manner. The rise of the crossover SUV will add to the challenge of finding a balanced approach.

Flexibility

In general, regulations that allow manufacturers flexibility in choosing how to achieve the desired policy goal (such as reducing fuel use or improving safety) are likely to lower the costs to the nation. That is because restricting the available technology options will reduce chances for cost-saving innovations. To the extent possible, consistent with the overall policy goals, flexibility is an important criterion. This is one reason for the committee's enthusiasm for tradeable fuel economy credits, as described in this chapter.

Hidden Costs of Forcing Technical Innovation

In general, it is risky to commercialize technologies while they are still advancing rapidly. Pioneering purchasers of vehicles that incorporate highly efficient new technology could find that these vehicles depreciate faster than those with old technology if the new technology results in higher repair costs or is replaced by improved versions. For the same reason, leasors might be reluctant to write leases on vehicles that are radically different and therefore have unpredictable future demand. GM offered its own leases on the EV1 (an electric vehicle) but did not sell the vehicle because it recognized that the car might be unusable in a few years because of technological obsolescence or high maintenance costs.

On the other hand, as explained in Chapter 3, foreign manufacturers are rapidly improving their technology, largely because their main markets are in countries with high fuel prices or high fuel economy standards. Not only are their vehicles economical (and frequently low in emissions), but they are proving popular because they offer other attributes valued by consumers, such as power and improved driving characteristics. Insofar as higher fuel economy standards force domestic manufacturers to adopt new technology, it could improve their competitiveness.

SAFETY IMPLICATIONS OF FUTURE INCREASES IN FUEL ECONOMY

In Chapter 2 the committee noted that the fuel economy improvement that occurred during the 1970s and early 1980s involved considerable downweighting and downsizing of the vehicle fleet. Although many general indicators of motor vehicle travel safety improved during that period (e.g., the fatality rate per vehicle mile traveled), the preponderance of evidence indicates that this downsizing of the vehicle fleet resulted in a hidden safety cost, namely, travel safety would have improved even more had vehicles not been downsized. Based on the most comprehensive and thorough analyses currently available, it was estimated in Chapter 2 that there would have been between 1,300 and 2,600 fewer crash deaths in 1993 had the average weight and size of the light-duty motor vehicle fleet in that year been like that of the mid-1970s. Similarly, it was estimated there would have been 13,000 to 26,000 fewer moderate to critical injuries. These are deaths and injuries that would have been prevented in larger, heavier vehicles, given the improvements in vehicle occupant protection and the travel environment that occurred during the intervening years. In other words, these deaths and injuries were one of the painful trade-offs that resulted from downweighting and downsizing and the resultant improved fuel economy.

This section of Chapter 4 addresses the question of how safety might be affected by future improvements in fuel economy. The key issue is the extent to which such improvements would involve the kind of vehicle downweighting and downsizing that occurred in the 1970s and 1980s. In Chapter 3 the committee examined the methods by which the industry can improve fuel economy in the future and identified many potential improvements in powertrains, aerodynamics, and vehicle accessories that could be used to increase fuel economy. Earlier in Chapter 4 the committee concluded that many of these technologies could pay for themselves in reduced fuel costs during the lifetimes of vehicles. Thus, it is technically feasible and potentially economical to improve fuel economy without reducing vehicle weight or size and, therefore, without significantly affecting the safety of motor vehicle travel. Two members of the committee believe that it may be possible to improve fuel economy without any implication for safety, even if downweighting is used. Their dissent forms Appendix A of this report.

The actual strategies chosen by manufacturers to improve fuel economy will depend on a variety of factors. Even if the technology included in the cost-efficient fuel economy improvement analyses is adopted, that technology might be used to provide customers with other vehicle attributes (performance, size, towing capacity) that they may value as much as or more than fuel economy. While it is clear vehicle weight reduction is not necessary for increasing fuel economy, it would be shortsighted to ignore the possibility that it might be part of the response to increases in CAFE standards.

In fact, in meetings with members of this committee, automotive manufacturers stated that significant increases in fuel economy requirements under the current CAFE system would be met, at least in part, by vehicle weight reduction. Because many automakers have already emphasized weight reduction with their current vehicle models, they also stated that substantial reductions in weight probably could not occur without some reduction in vehicle size. They were referring to exterior dimensions, not to interior space, which is a high customer priority. However, this type of size reduction also reduces those portions of the vehicle that provide the protective crush zones required to effectively manage crash energy.

When asked about the potential use of lighter materials to allow weight reduction without safety-related size reductions, the manufacturers acknowledged that they were gaining more experience with new materials but expressed concern about their higher costs. Given that concern, industry representatives did not expect that they could avoid reducing vehicle size if substantial reductions in vehicle weight were made. Thus, based on what the committee was told in direct response to its questions, significant increases in fuel economy requirements under the current CAFE system could be accompanied by reductions in vehicle weight, and at some level, in vehicle size.

The committee recognizes that automakers' responses could be biased in this regard, but the extensive downweighting and downsizing that occurred after fuel economy requirements were established in the 1970s (see Chapter 2) suggest that the likelihood of a similar response to further increases in fuel economy requirements must be considered seriously. Any reduction in vehicle size and weight would have safety implications. As explained in Chapter 2, there is uncertainty in quantifying these implications. In addition, the societal effects of downsizing and downweighting depend on which segments of the fleet are affected. For example, if future weight reductions occur in only the heaviest of the light-duty vehicles, that can produce overall improvements in vehicle safety. The following sections of the report describe the committee's findings on vehicle weight, size, and safety in greater detail and set forth some general concerns and recommendations with regard to future efforts to improve the fuel economy of the passenger vehicle fleet.

The Role of Vehicle Mass

The 1992 NRC fuel economy report concluded as follows: "Although the data and analyses are not definitive, the Committee believes that there is likely to be a safety cost if downweighting is used to improve fuel economy (all else being equal)" (NRC, 1992, p. 6). Studies continue to accumulate indicating that mass is a critical factor in the injury outcomes of motor vehicle crashes (e.g., Evans and Frick, 1992, 1993; Wood, 1997; Evans, 2001). Although there are arguments that not all increases in vehicle weight benefit safety, these arguments do not contradict the general finding that, all other things being equal, more mass is protective. For example, Joksch et al. (1998) have reported that, when vehicles of similar size are compared, those that are significantly heavier than the average for their size do not appear to improve their occupants' protection but do increase the risk to occupants of other vehicles with which they collide. The authors note (Joksch et al., 1998, p. ES-2) that this effect should be interpreted with caution, because it is likely that overweight vehicles are overweight in part because of more powerful (and heavier) engines and performance packages, which could attract more aggressive drivers. Thus, Joksch et al. demonstrate that some weight increases can be detrimental to safety, but this finding does not change the basic relationship: among vehicles of equal size and similar driving exposure, the heavier vehicle would be expected to provide greater occupant protection.

The protective benefits of mass are clearly understood in multiple-vehicle crashes, where the physical conservation of momentum results in the heavier vehicle's experiencing a smaller change in momentum, and hence lower occupant deceleration, than the lighter vehicle. However, mass is also protective in single-vehicle crashes with objects (such as trees, poles, or guard rails), because many of these objects will move or deform in proportion to the mass of the vehicle. In this case, the change in velocity is not affected, but the deceleration of the vehicle and its occupants decreases as the object that is struck deforms or moves. Figures 4-7 and 4-8, which show fatality rates per million registered vehicles by vehicle type and vehicle weight, illustrate the protective effects of vehicle mass in single- and multiple-vehicle crashes. The heaviest vehicles in each class (cars, SUVs, and pickups) have about half as many fatalities per registered vehicle as the lightest vehicles.[1]

While the benefits of mass for self-protection are clear, mass can also impose a safety cost on other road users. In a collision between two vehicles, increasing the mass of one of the vehicles will decrease its momentum change (and the forces on its occupants) but increase the momentum change of the crash partner. Figure 4-9 shows the increased fatalities caused in other vehicles per million registered cars, SUVs, or pickups as the mass of the "striking" vehicle increases. Because of this tendency to cause more injuries to occupants of other vehicles, heavier vehicles are sometimes said to be more "aggressive." Aggressivity also varies by vehicle type; SUVs and pickups cause more deaths in other vehicles than do passenger cars. There are also effects of vehicle type on the likelihood of injuring pedestrians and other vulnerable road users, although the effects of mass are much weaker there (see Figure 4-10). The smaller effect of mass for the more vulnerable nonoccupants probably reflects the fact that mass ratios between, for example, pedestrians and the lightest vehicle are already so high that increasing mass of the vehicle makes little additional difference in survivability for the pedestrian. The differences by vehicle type may reflect

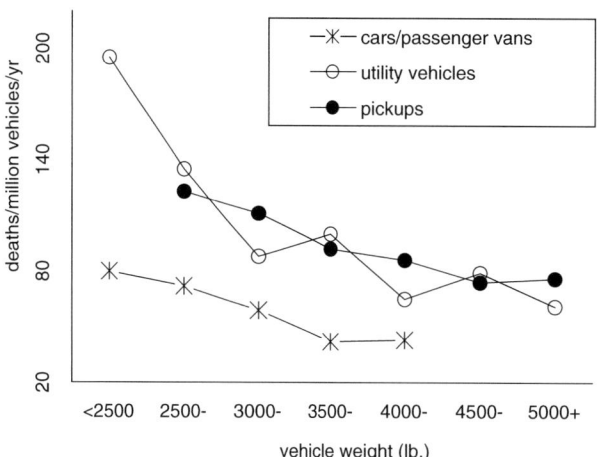

FIGURE 4-7 Occupant death rates in single-vehicle crashes for 1990–1996 model passenger vehicles by weight of vehicle. SOURCE: Insurance Institute for Highway Safety, using fatality data from NHTSA's Fatality Analysis Reporting System (FARS) for 1991–1997, a census of traffic fatalities maintained by NHTSA, and vehicle registration data from the R.L. Polk Company for the same years.

the greater propensity of taller vehicles to do more damage in collisions with nonoccupants.

The net societal safety impact of a change in the average mass of the light-duty vehicle fleet can be an increase, a decrease, or no change at all. The outcome depends on how that change in mass is distributed among the vehicles that

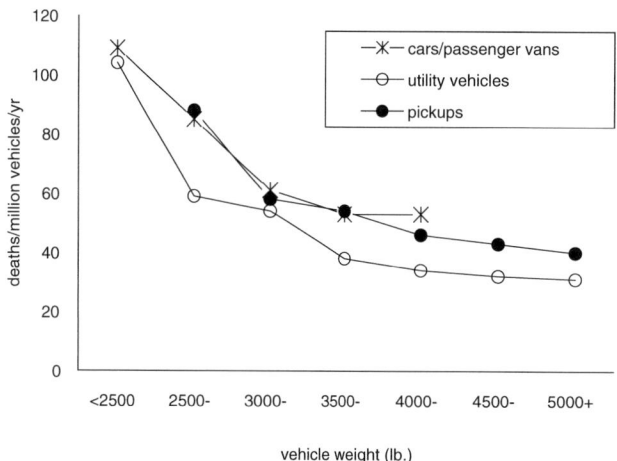

FIGURE 4-8 Occupant death rates in two-vehicle crashes for 1990–1996 model passenger vehicles by weight of vehicle. SOURCE: Insurance Institute for Highway Safety, using fatality data from NHTSA's Fatality Analysis Reporting System (FARS) for 1991–1997, a census of traffic fatalities maintained by NHTSA, and vehicle registration data from the R.L. Polk Company for the same years.

---

[1] Because mass and size are highly correlated, some of the relationships illustrated in these figures are also attributable to differences in vehicle size, an issue that will be discussed further later in this chapter. These figures also show that mass is not the only vehicle characteristic affecting occupant injury risk, as there are substantial differences in occupant fatality risk for cars, SUVs, and pickups of similar weight. Much of this difference probably results from the higher ride heights of SUVs and pickups; riding higher is protective in multiple-vehicle crashes, because the higher vehicle tends to override lower vehicles, but increases single-vehicle crash fatalities by raising the vehicle's center of gravity and increasing rollover risk.

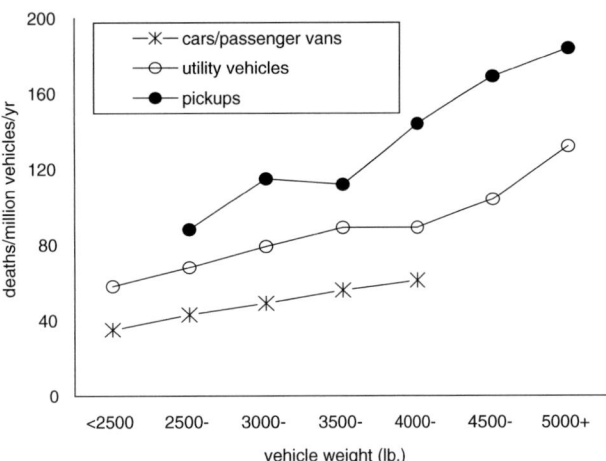

FIGURE 4-9 Occupant death rates in other vehicles in two-vehicle crashes for 1990–1996 model passenger vehicles. SOURCE: Insurance Institute for Highway Safety, using fatality data from NHTSA's Fatality Analysis Reporting System (FARS) for 1991–1997, a census of traffic fatalities maintained by NHTSA, and vehicle registration data from the R.L. Polk Company for the same years.

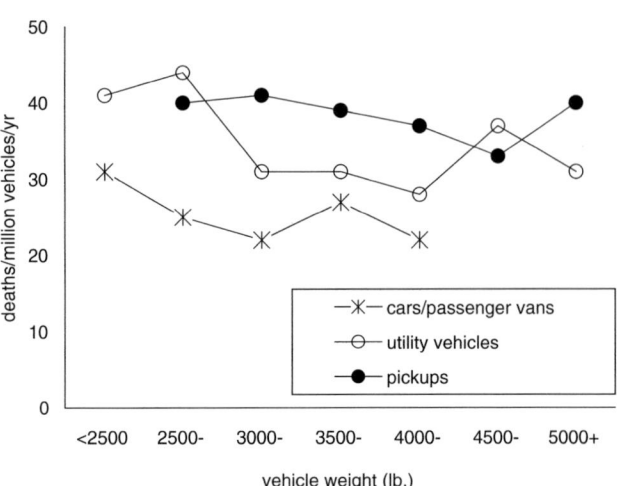

FIGURE 4-10 Pedestrian/bicyclist/motorcyclist death rates for 1990–1996 model passenger vehicles by vehicle weight. SOURCE: Insurance Institute for Highway Safety, using fatality data from NHTSA's Fatality Analysis Reporting System (FARS) for 1991–1997, a census of traffic fatalities maintained by NHTSA, and vehicle registration data from the R.L. Polk Company for the same years.

make up the vehicle fleet (compare Appendix D of NRC, 1992).

In Chapter 2 the committee reviewed a 1997 study of the National Highway Traffic Safety Administration (NHTSA) (Kahane, 1997) that estimated the anticipated effect of a 100-lb change in the average weight of the passenger car and light-truck fleets. These estimates indicated that such a weight reduction would have different effects on different crash types. Nevertheless, over all crash types, decreasing the average weight of passenger cars would be expected to increase the motor vehicle fatality risk (all else being equal). Correspondingly, decreasing the average weight of the heavier fleet of light trucks might reduce the motor vehicle crash fatality risk (although the latter result was not statistically significant).

Lund and Chapline (1999) have reported similar findings. They found that total fatalities in a hypothetical fleet of relatively modern passenger vehicles would be reduced by about 0.26 percent if all pickups and SUVs weighing more than 4,000 lb were replaced with pickups and SUVs weighing 3,500 to 4,000 lb. However, if the heaviest cars, those weighing more than 3,500 lb, were replaced by cars weighing between 3,000 and 3,500 lb, the estimated effect changed to an increase in total fatalities of 4.8 percent. Although the authors did not consider crashes with nonoccupants such as pedestrians and bicyclists, whose risk appeared to have no clear relationship to vehicle mass (see Figure 4-10), the results confirm Kahane's (1997) finding that the expected societal effect of decreases in vehicle mass, whether in response to fuel economy requirements or other factors, depends on which part of the vehicle fleet becomes lighter.[2]

---

[2]As noted in Chapter 2, the committee has relied heavily on these NHTSA analyses in its consideration of the likely effects of future improvements in fuel economy. Although they have been the subject of controversy, the committee's own review of the analyses found no compelling reasons to change the conclusions or alter the estimates of the relationship between vehicle mass and safety. Statistical and conceptual considerations indicate that changes are occurring in the pattern of motor vehicle crash fatalities and injuries, which could affect these estimates in the future, but the stability of relative occupant fatality rates between lighter and heavier vehicles over the past 20 years does not suggest that the estimates should change significantly over the time period considered by the committee, essentially the next 10–15 years (see Chapter 2). As indicated in Chapter 2, this committee believes that further study of the relationship between size, weight, and safety is warranted, because of uncertainty about the relationship between vehicle weight and safety. However, the majority of the committee believes that these concerns should not prevent the use of NHTSA's careful analyses to provide some understanding of the likely effects of future improvements in fuel economy, if those improvements involve vehicle downsizing. The committee believes this position is consistent with that of the National Research Council's Transportation Research Board (NRC-TRB) committee that reviewed the draft NHTSA report in 1996, which concluded "it is important . . . to provide a sense of the range of uncertainty so that policy makers and researchers can properly interpret the results" (NRC-TRB, 1996, p. 7). NHTSA's final report has done that, and the use of NHTSA's results is consistent with this committee's approach to other uncertainties surrounding efforts to improve fuel economy: that is, to use the best scientific evidence available to gauge likely effects and to state the uncertainty associated with those efforts.

Vehicle Size and Safety

Estimates of the effect of vehicle weight have been confounded with the effects of size because the mass and size of vehicles are correlated so closely. For example, the 2001 Buick LeSabre, a typical large car, is 200 inches long and 74 inches wide and weighs (unloaded) 3,600 lb. A 2001 four-door Honda Civic, a typical small car for the United States, has an overall vehicle length of 175 inches, overall width of 67 inches, and a weight of 2,400 lb (Highway Loss Data Institute, 2001). Thus, some of the negative effect of vehicle weight constraints on safety that has been attributed to mass reduction is attributable to size reduction.

Historical changes in the fleet have similarly confounded mass and size characteristics. While the mass of the passenger car fleet was decreased about 900 pounds between 1975 and 1990, the length of vehicles also declined, with average wheelbase (the distance between the front and rear axles) declining more than 9 inches (NRC, 1992). Efforts to disentangle these effects are hampered by the fact that, when vehicles of similar size differ in mass, they usually do so for reasons that still confound the estimate of mass effects (see, for example, Joksch et al., 1998; C.J. Kahane, NHTSA, also spoke of this in his presentation to the committee on February 6, 2001).

Despite this confounding, carefully controlled research has demonstrated that, given a crash, larger vehicles provide more occupant protection independent of mass. In crashes between vehicles of similar mass, smaller vehicles have higher fatality rates than larger ones (Evans and Frick, 1992; Wood, 1997; Evans, 2001). In addition, Wood (1997) has argued that much of the apparent protective effect of mass in single-vehicle crashes may occur because of the association of mass with size. Theoretically, increased size of one vehicle can be beneficial to other road users as well, to the extent that the increased size translates to more crush space (Ross and Wenzel, 2001; O'Neill, 1998). By the same token, to the extent that size reductions translate to less crush space, smaller size is detrimental to both the vehicle's own occupants and the occupants of other vehicles.

Theoretically, size can affect crash likelihood as well as crashworthiness, independently of mass. For example, reductions in size may make a vehicle more maneuverable. Smaller vehicles may be easier to "miss" in the event of a potential collision. Drivers of smaller, lower vehicles may be better able to see and avoid other vulnerable road users such as pedestrians and bicyclists. These effects could lead to reduced societal risk despite the negative effects of reduced size on occupant protection, given a crash.

The committee found no direct empirical support for this crash avoidance benefit, but there is some indirect support in the pattern of Kahane's (1997) fatality results. For example, Kahane reported a much larger reduction in pedestrian, bicyclist, and motorcyclist fatalities associated with a reduction in light truck weights than with the same reduction in car weights. The mass ratios between pedestrians and motor vehicles are so large that the difference between cars and light trucks cannot be due solely to the change in mass. Rather, it is possible that some of the difference in reduction is due to changes in size, and hence visibility of pedestrians, to the driver. In addition, Kahane's analyses found no expected change in car-to-car or truck-to-truck crash fatalities as a result of vehicle downweighting. That result is contradictory to other studies indicating that, in the event of a crash, smaller cars offer less occupant protection (see above). The explanation could be that a smaller vehicle's increased injury risk given a crash is offset by a tendency to get into fewer crashes (by virtue of its easier-to-miss, smaller profile or its potentially superior agility).

The direct evidence that is available contradicts this crash avoidance hypothesis, however. Overall, there is evidence that smaller vehicles actually are involved in more collisions than larger vehicles, relative to their representation in the population of vehicles. The Highway Loss Data Institute (HLDI) tracks the collision insurance claims experience of about 65 percent of the vehicles insured in the United States. Table 4-4 indicates the incidence of collision claims for 1998–2000 models in 1998–2000, relative to the average collision claims frequency for all passenger cars. The results have been standardized to represent similar proportions of drivers less than 25 years of age. The principal pattern of these data is that the frequency of collision claims is higher for smaller vehicles, the opposite of what might be expected from the simple geometry of the vehicles.

The important theoretical role of vehicle size for crashworthiness led the committee to consider whether some of the adverse safety effects associated with decreases in mass could be mitigated if size remained the same. In this context, it is important to distinguish among different meanings of the term "vehicle size." Changes in size of the occupant compartment are generally less relevant to safety (but not irrelevant) than changes in the exterior size of vehicles—the latter affect the size of the so-called crush zones of the vehicle.

It is noteworthy that almost all of the downsizing that occurred in conjunction with increased fuel economy occurred in this crush zone; the size of the occupant compart-

TABLE 4-4 Relative Collision Claim Frequencies for 1998–2000 Models

Size Class	Four-Door Cars	SUVs	Pickups
Mini	124	—	—
Small	112	78	87
Midsize	99	76	—
Large	88	65	75
Very large	70	86	—

SOURCE: Highway Loss Data Institute (2001).

ments of vehicles has changed little since 1977, when EPA estimated the interior volume of cars to be 110 cubic feet. That fell to a low of 104 cubic feet in 1980, but generally ranged from 108 to 110 cubic feet during the 1990s (EPA, 2000). This reflects the critical effect of interior volume on the utility of passenger cars and hence their marketability. For the same reason, the committee expects that future size reduction, if necessary to reduce weight, would again occur principally outside the occupant compartment, and that interior space would be one of the last areas subject to reduction.

What would be the benefits if crush space were retained in a future lighter-weight fleet? Empirical data do not exist to answer this question quantitatively, but the committee would expect a smaller adverse safety effect of vehicle weight reduction. Still there would be some loss of occupant protection, because lighter vehicles decelerate more rapidly in crashes with other vehicles or with deformable fixed objects. Effective crush space would therefore have to increase with reductions in mass in order to keep injury risk the same. Thus, if manufacturers try to maintain current levels of occupant protection as they downweight, they would have to use some of the weight savings from alternative materials and structures to provide additional crush space.

In addition, it must be noted that not all the effects of maintaining size will necessarily be beneficial. Some of the increased risk of injury to occupants associated with vehicle downweighting was offset by reduced injury risk to vulnerable road users (Kahane, 1997). Those offsetting benefits may occur as a result of changes in certain size characteristics that are typically associated with weight reduction. For example, a change in the height and size of vehicle front ends may be the critical factor in the estimated benefits to pedestrians of reductions in the weight of light trucks. Thus, if size is maintained, it may reduce the benefits of lighter-weight vehicles for pedestrians and cyclists.

In sum, the committee believes that it will be important to maintain vehicle crush space if vehicles are downweighted in the future, but it is unable to develop quantitative estimates of the extent to which such efforts can reduce the estimated effects of vehicle downweighting.

Effect of Downweighting by Vehicle and Crash Type

To examine the impact of vehicle weight reduction on the future safety environment, the majority of the committee considered how weight reduction might influence various crash types. This section of the report discusses how the committee expects vehicle weight reduction to operate in various kinds of multivehicle and single-vehicle crashes.

Multiple-Vehicle Crashes

Multiple-vehicle crashes account for slightly more than half of occupant fatalities. They include crashes between cars and light trucks, crashes between either cars or light trucks and heavy trucks, and crashes of cars with cars and light trucks with light trucks. The effect on safety of downweighting and downsizing of the light-duty fleet varies among these multiple-vehicle crash types:

Collisions Between Cars and Light Trucks Uniform downweighting of the fleet, that is, reducing weight from all classes of vehicles, might be expected to produce an increase in traffic casualties. However, if the downweighting is restricted to the heaviest pickup trucks and SUVs, with no weight reduction in the smaller light trucks or passenger cars, casualties could decrease. Thus, for these crashes, any change in casualties is very sensitive to how downsizing is distributed. This is a consistent finding in the safety literature and is confirmed by the 1997 NHTSA analysis.

Collisions Between Light-Duty Vehicles and Heavy Trucks If there is downsizing in any light-duty vehicles with no corresponding change in heavy trucks, the number of casualties would increase. This finding is reflected in the 1997 NHTSA analyses and throughout the safety literature.

Collisions of Cars with Cars or Trucks with Trucks In collisions of like vehicles of similar weight, a reduction in both vehicle weight and size is expected to produce a small increase in casualties. In fact, one consistent finding in the safety literature is that as average vehicle weight declines, crash risks increase (see Evans [2001], for example). The laws of momentum do not explain this typical finding, but analysts typically attribute it to the fact that as vehicle weight declines, so does vehicle size. Narrowly defined studies focusing on pure crashworthiness consistently show this increase. However, the 1997 NHTSA analysis, which examined the entire traffic environment, predicted a small (statistically insignificant) decrease in fatalities. It has been hypothesized that this inconsistency might be explained by the greater maneuverability of smaller vehicles and, hence, their involvement in fewer crashes. Studies have not found this to be the case, but it is very difficult to fully normalize data to account for the very complex traffic environment. In this instance, the committee is unable to explain why the 1997 NHTSA analysis does not produce results consistent with those in the safety literature. The rest of NHTSA's findings regarding multivehicle crash types are consistent with the literature.

Single-Vehicle Crashes

Single-vehicle crashes account for almost half of light-duty vehicle occupant fatalities. As with multiple-vehicle crashes, the safety literature indicates that as vehicle weight and vehicle size decline, crash risks increase. The committee examined three types of single-vehicle crashes: rollovers, crashes into fixed objects, and crashes with pedestrians and bicyclists.

Rollovers Historically, a large part of the increase in casualties associated with vehicle downsizing has been attributable to rollovers. As vehicles get shorter and narrower they become less stable and their rollover propensity increases. If the length and width of the vehicle can be retained as weight is removed, the effect of weight reduction on rollover propensity can be reduced. However, unless track width widens, the vehicle's rollover propensity in actual use would still increase somewhat, because occupants and cargo are typically located above the vehicle's center of gravity (CG), raising the vehicle's CG-height in use.

One Ford safety engineer noted in discussions with the committee that the application of crash avoidance technology (generically known as electronic stability control) that is being introduced on some new vehicles might significantly reduce vehicle rollovers. It is unknown at this time how effective this technology will be. In this regard, it is worthwhile noting the experience with other technology aimed at reducing crash likelihood.

Antilock brakes were introduced on vehicles after extensive testing indicated they unequivocally improved vehicle handling in emergency situations. In fact, antilock brakes were cited in the 1992 NRC report as a new technology that might offset the negative effects of vehicle mass reductions caused by increased fuel economy requirements (NRC, 1992, p. 59). However, experience with antilock brakes on the road has been disappointing. To date, there is no evidence that antilock brakes have affected overall crash rates at all; the principal effect has been to change the pattern of crashes (Farmer et al., 1997; Farmer, 2001; Hertz et al., 1998). Most tellingly, the initial experience suggested that antilock brakes actually increased fatal, single-vehicle, run-off-the-road crashes and rollovers (Farmer et al., 1997), the very crashes that the NRC committee thought antilock brakes might benefit. Recent research suggests that this increase in fatal, single-vehicle crashes associated with antilocks may be diminishing (Farmer, 2001), but still the evidence does not permit a conclusion that antilocks are reducing crash risk.

Thus while it is conceivable that new technology might reduce or eliminate the risk of rollovers, at this time the committee is not willing to change its expectations based on this eventuality.

Fixed-Object Collisions Fixed-object collisions occur with both rigid, unyielding objects and yielding objects that can break or deform. For crashes into rigid objects, larger vehicles will, on average, permit longer, lower decelerations, although this will be influenced by the stiffness of the vehicle. In crashes with yielding, deformable objects, as a vehicle's weight is reduced, the struck object's deformation is reduced, increasing the deceleration in the striking vehicle. Thus lighter vehicles, with or without size reductions, would be expected to increase occupant fatality risk.

Studies by Evans (1994), Klein et al. (1991), and Partyka and Boehly (1989) all confirm that there is an inverse relationship between occupant safety and vehicle weight in these crashes. Klein et al. found that there was a 10 percent increase in fatality risk associated with a 1,000-lb reduction in vehicle weight in single-vehicle, nonrollover crashes. Similarly, NHTSA's 1997 analysis estimated that there is a slightly greater than 1 percent increase in fatality risk associated with a 100-lb reduction in vehicle weight in these crashes.[3]

Collisions with Pedestrians, Bicyclists, and Motorcyclists
If vehicles are downsized, they may be less likely to strike a pedestrian, bicyclist, or motorcyclist. Further, should a collision occur, the reduced mass could lead to a reduction in casualties, although, given the mismatch in the mass between the vehicle and the unprotected road user, any benefit would not be expected to be very large.

Safety Impacts of Possible Future Fuel Economy Scenarios

The foregoing discussion provides the majority of the committee's general views on how vehicle weight and size reductions, if they occur in response to demands for increased fuel economy, could affect the safety of motor vehicle travel. The committee evaluated the likely safety effects of a number of possible scenarios involving different amounts of downweighting of the future fleet, using the results of the NHTSA analyses (Kahane, 1997) for quantitative guidance. Two of these possible scenarios are presented below. They are not intended to reflect the committee's recommendations, but they do reflect two possibilities that can be instructive in evaluating future fuel economy requirements. The first scenario examines the safety consequences that would be expected if manufacturers achieved a 10 percent improvement in fuel economy using the same pattern of

---

[3]In an attempt to gain more insight into single-vehicle crashes, Partyka (1995) examined field data on single-vehicle, nonrollover crashes into yielding fixed objects. Her goal was to learn if there was a relationship between vehicle weight and crash outcomes. However, in this study, rather than examining occupant injury as the crash outcome of interest, crash outcomes were defined as whether the yielding object was damaged in the crash. For frontal crashes she found that damage to the yielding fixed object was more likely to occur in crashes with heavier vehicles. However, there was no consistent relationship between vehicle weight and damage to the yielding object in side crashes. The author was not able to consider crash severity as a possible explanation for this anomaly since NHTSA's national accident sampling system (NASS) did not provide estimates of crash severity for any of the events where the yielding object was damaged. Further, the author did not aggregate the data to reach an overall conclusion regarding the relationship between vehicle weight and damage to yielding fixed objects because she was uncomfortable with the results of the side impact analysis and questioned its reliability (personal communication from the author).

downweighting and downsizing that occurred in the late 1970s and 1980s. As discussed above, it is uncertain to what extent downweighting and downsizing will be part of manufacturers' strategies for improving fuel economy in the future, but one guide to their behavior can be to look at what they did in the past.

The second scenario is based on the minimal weight change predictions associated with the committee's cost-efficient analysis of future technology, discussed earlier in this chapter. Thus, this analysis projects the safety consequences if manufacturers apply likely powertrain technology primarily to improve vehicle fuel economy.

In both scenarios, the hypothetical weight reductions are based on the MY 2000 light-duty vehicle fleet (the average car weighed 3,386 lb and the average light truck, 4,432 lb). The estimated effects of the weight reductions are applied to the 1993 vehicle fleet, the reference year for the NHTSA analyses (Kahane, 1997). Obviously the results in these scenarios cannot be, nor are they intended to be, precise projections. NHTSA's analysis was based on the characteristics of 1985–1993 vehicles and was applied to the 1993 fleet. Further, even if manufacturers dropped the average weight of the vehicles described in the two scenarios starting in 2002, it would take many years before those changes had any significant influence on the overall makeup of the future vehicle fleet. And that fleet would be operating in an environment that will differ from the environment that was the basis for the 1997 NHTSA analysis.

Nevertheless, the committee's majority believes that the calculations below provide some insight into the safety impacts that might occur if manufacturers reduce weight in response to increased fuel economy standards, as described below.

Historical Pattern Scenario

Between 1975 and 1984, the average weight of a new car dropped from 4,057 to 3,098 lbs (24 percent). For light trucks, weight declined from 4,072 to 3,782, or 290 lb (7 percent). These mass reductions suggest a 17 percent improvement in fuel economy for cars and a 5 percent improvement for light trucks.[4] During those years, the fuel economy of cars and light trucks actually improved by about 66 percent (from 15.8 mpg to 26.3 mpg) and 50 percent (from 13.7 mpg to 20.5 mpg), respectively. In other words, vehicle downsizing accounted for about 25 percent of the improvement in fuel economy of cars and 10 percent of the improvement of light trucks between 1975 and 1984.

A similar pattern in achieving a 10 percent improvement in fuel economy today would imply a 3.6 percent weight reduction for cars (122 lb for 2000 models) and a 1.4 percent weight reduction for light trucks (63 lb for 2000 models). Had this weight reduction been imposed on the fleet in 1993, it would have been expected to increase fatalities involving cars about 370 (±110) and to decrease fatalities involving light trucks by about 25 (±40). The net effect in 1993 for a 10 percent improvement in fuel economy with this mix of downsizing and increased fuel efficiency would have been about 350 additional fatalities (95 percent confidence interval of 229 to 457).

Cost-Efficient Scenario

Earlier in this chapter (under "Analysis of Cost-Efficient Fuel Economy"), cost-efficient fuel economy increases of 12 to 27 percent for cars and 25 to 42 percent for light trucks were estimated to be possible without any loss of current performance characteristics. In the cost-efficient analysis, most vehicle groups actually gain approximately 5 percent in weight to account for equipment needed to satisfy future safety standards. Thus, for these vehicle groups, cost-efficient fuel economy increases occur without degradation of safety. In fact, they should provide enhanced levels of occupant protection because of both the increased level of safety technology and the increased weight of that technology.

For three groups of vehicles (large cars, midsize SUVs, and large SUVs), under the cost-efficient scenario, it is projected that manufacturers would probably compensate for the added weight attributable to safety technology with weight reductions in other areas. These weight reductions mean that occupants of these vehicles are somewhat less protected in crashes than they otherwise would have been (though they still benefit from the added safety technology).

Estimating the potential effect of this limited downsizing is complicated because the estimates of the safety effect of 100-lb changes in average vehicle weight developed by NHTSA cannot be applied directly to changes in specific vehicle groups. However, NHTSA's report included sensitivity analyses that estimated the effect of weight reductions restricted to the heaviest 20 percent of cars (those heavier than 3,262 lb) and to the heaviest 20 percent of light trucks (those heavier than 3,909 lb) (Kahane, 1997, pp. 165–172). Specifically, NHTSA estimated that a 500-lb reduction in these vehicle groups in 1993 would have resulted in about 250 additional fatalities in car crashes and about 65 fewer fatalities in light-duty truck crashes.

If it is assumed that cars greater than 3,262 lb correspond roughly to "large cars" as referred to in the committee's cost-efficient fuel economy analysis outlined earlier in this chapter, then the 5 percent weight reduction, had it occurred in 1993, would have been about 160 to 180 lb, or only about one-third of the reduction of the NHTSA analysis. Thus, it is estimated that reducing the weight of large cars in the way foreseen in the cost-effective fuel economy analyses would have produced about 80 additional fatalities in car crashes in 1993.

---

[4]Based on an assumed 7 percent improvement in fuel economy for each 10 percent reduction in weight.

Similarly, if midsize and large SUVs are assumed to be in the heaviest 20 percent of light-duty trucks and if such vehicles account for as much as half the light-duty truck population in the near future, then about half the light-duty truck fleet will be 5 percent lighter (about 200 lb). This weight reduction is about 40 percent of the reduction studied by NHTSA and would have saved 15 lives had it occurred in 1993.

Adding these effects yields an estimated increase of 80 car crash fatalities minus 15 fewer truck crash fatalities, or about 65 additional fatalities.

NHTSA's sensitivity analysis did not include standard errors for the weight reductions in the heaviest vehicles, but in this case they would clearly be large enough that reasonable confidence bounds for this estimate would include zero. Thus, an increase in fatality risk in the future fleet is predicted from the kinds of weight reductions included in the cost-efficient fuel economy scenario identified by the committee, but the uncertainty of this estimate is such that the effect might be zero and would be expected to result in fewer than 100 additional fatalities.

In addition, it should be noted that these results do not imply that the actual safety effect would be as small as this if the required fuel economy rises to the targets indicated in the cost-efficient analyses. This is because manufacturers could choose to use advances in drive train technology for other vehicle attributes such as acceleration or load capacity. In that case, additional vehicle downweighting might occur, and the adverse safety consequences would grow. Thus, the actual safety implications of increasing fuel economy to the cost-efficient levels specified earlier in this chapter will depend on what strategies manufacturers actually choose in order to meet them and the structure of the regulatory framework.

Conclusion: Safety Implications of Increased CAFE Requirements

In summary, the majority of the committee finds that the downsizing and weight reduction that occurred in the late 1970s and early 1980s most likely produced between 1,300 and 2,600 crash fatalities and between 13,000 and 26,000 serious injuries in 1993. The proportion of these casualties attributable to CAFE standards is uncertain. It is not clear that significant weight reduction can be achieved in the future without some downsizing, and similar downsizing would be expected to produce similar results. Even if weight reduction occurred without any downsizing, casualties would be expected to increase. Thus, any increase in CAFE as currently structured could produce additional road casualties, unless it is specifically targeted at the largest, heaviest light trucks.

For fuel economy regulations not to have an adverse impact on safety, they must be implemented using more fuel-efficient technology. Current CAFE requirements are neutral with regard to whether fuel economy is improved by increasing efficiency or by decreasing vehicle weight. One way to reduce the adverse impact on safety would be to establish fuel economy requirements as a function of vehicle attributes, particularly vehicle weight (see Chapter 5). Another strategy might be to limit horsepower-to-weight ratios, which could save fuel by encouraging the application of new fuel efficiency technology for fuel economy rather than performance.

The committee would also like to note that there is a remarkable absence of information and discussion of the speed limit in the context of fuel consumption. The national 55-mph speed limit that was in effect until 1987 is estimated to have reduced fuel consumption by 1 to 2 percent while simultaneously preventing 2,000 to 4,000 motor vehicle crash deaths annually (NRC, 1984). Relaxation of the speed limit has increased fatalities in motor vehicle crashes (Baum et al., 1991; Farmer et al., 1999), and fuel consumption presumably is again higher than it would have been. In addition, it is reasonable to expect that higher speed limits, combined with higher fuel efficiency, provide incentives for driving further and faster, thereby offsetting intended increases in fuel economy, though this hypothesis is speculative on the part of the committee.

But the committee wants to close where it started, by considering the effect of future advances in safety technology on CAFE. Despite the adverse safety effects expected if downweighting occurs, the effects are likely to be hidden by the generally increasing safety of the light-duty vehicle fleet. This increase in safety is driven by new safety regulations, testing programs that give consumers information about the relative crash protection offered by different vehicles, and better understanding of the ways in which people are injured in motor vehicle crashes. Alcohol-impaired driving has been decreasing and seat belt usage has been increasing. Roads are becoming less hazardous for vehicles.

Some might argue that this improving safety picture means that there is room to improve fuel economy without adverse safety consequences. However, such a measure would not achieve the goal of avoiding the adverse safety consequences of fuel economy increases. Rather, the safety penalty imposed by increased fuel economy (if weight reduction is one of the measures) will be more difficult to identify in light of the continuing improvement in traffic safety. Just because these anticipated safety innovations will improve the safety of vehicles of all sizes does not mean that downsizing to achieve fuel economy improvements will have no safety costs.

If an increase in fuel economy is effected by a system that encourages either downweighting or the production and sale of more small cars, some additional traffic fatalities would be expected. Without a thoughtful restructuring of the program, that would be the trade-off that must be made if CAFE standards are increased by any significant amount.

REFERENCES

Austin, T.C., R.G. Dulla, and T.R. Carlson. 1999. Alternative and Future Technologies for Reducing Greenhouse Gas Emissions from Road Vehicles. Prepared for the Committee on Road Vehicle Technology and Fuels, Natural Resources Canada. Sacramento, Calif.: Sierra Research, Inc., July.

Baum, H.M., J.K. Wells, and A.K. Lund. 1991. "The Fatality Consequences of the 65 mph Speed Limits, 1989." *Journal of Safety Research*, 22: 171–177.

Butters, Jamie. 2001. "Outlook Lowered for Ford and GM: S&P Expresses Concern About Creditworthiness." *Detroit Free Press*, February 7.

Charles River Associates, Inc. 1995. Policy Alternatives for Reducing Petroleum Use and Greenhouse Gas Emissions: The Impact of Raising Corporate Average Fuel Economy (CAFE) Standards, July.

DeCicco, J., and D. Gordon. 1993. Steering with Prices: Fuel and Vehicle Taxation as Market Incentives for Higher Fuel Economy, December. Washington, D.C.: ACEEE.

DeCicco, J., F. An, and M. Ross. 2001. Technical Options for Improving the Fuel Economy of U.S. Cars and Light Trucks by 2010–2015, April. Washington, D.C.: American Council for an Energy-Efficient Economy (ACEEE).

EEA (Energy and Environmental Analysis, Inc.). 2001. Technology and Cost of Future Fuel Economy Improvements for Light-Duty Vehicles. Draft Final Report. Prepared for the committee and available in the National Academies' public access file for the committee's study.

EIA (Energy Information Administration). 2001. Annual Energy Outlook 2001. Table A 7, Transportation Sector Key Indicators and Delivered Energy Consumption. Washington, D.C.: EIA.

EPA (Environmental Protection Agency). 2000. Light-Duty Automotive Technology and Fuel Economy Trends 1975 Through 2000. EPA-R-00-008. December 2000.

Evans, L. 1994. "Driver and Fatality Risk in Two-Car Crashes Versus Mass Ratio Inferred." *Accident Analysis and Prevention* 26: 609–616.

Evans, L. 2001. "Causal Influence of Car Mass and Size on Driver Fatality Risk." *American Journal of Public Health* 91(7): 1076–1081.

Evans, L., and M.C. Frick. 1992. "Car Size or Car Mass: Which Has Greater Influence on Fatality Risk." *American Journal of Public Health* 82 (8): 1105–1112.

Evans, L., and M.C. Frick. 1993. "Mass Ratio and Relative Driver Fatality Risk in Two-Vehicle Crashes." *Accident Analysis and Prevention* 25 (2): 213–224.

Farmer, C.M. 2001. New Evidence Concerning Fatal Crashes of Passenger Vehicles Before and After Adding Antilock Braking Systems. *Accident Analysis and Prevention* 33: 361–369.

Farmer, C.M., A. Lund, R. Trempel, and E. Braver. 1997. "Fatal Crashes of Passenger Vehicles Before and After Adding Antilock Braking Systems." *Accident Analysis and Prevention* 29: 745–757.

Farmer, C.M., R.A. Retting, and A.K. Lund. 1999. "Changes in Motor Vehicle Occupant Fatalities After Repeal of the National Maximum Speed Limit." *Accident Analysis and Prevention* 31: 537–543.

Greene, D.L. 2001. Cost-Efficient Economy Analysis. Draft. Prepared for the committee and available in the National Academies' public access file for the committee's study.

Greene, D.L., and Steven E. Plotkin. 2001. Energy Futures for the U. S. Transportation Sector. Draft, April.

Greene, D.L., and J. DeCicco. 2000. Engineering-Economic Analyses of Automotive Fuel Economy Potential in the United States. ORNL/TM–2000/26, February. Oak Ridge, Tenn.: Oak Ridge National Laboratory.

Hertz, E., J. Hilton, and D. Johnson. 1998. Analysis of the Crash Experience of Vehicles Equipped with Antilock Braking Systems—An Update (98-S2-O-07). Proceedings of the 16th International Technical Conference on the Enhanced Safety of Vehicles. Washington, D.C.: NHTSA.

Highway Loss Data Institute. 2001. Fatality Facts: Passenger Vehicles. Available online at <http://www.carsafety.org>.

Joksch, H., D. Massie, and R. Pichler. 1998. Vehicle Aggressivity: Fleet Characterization Using Traffic Collision Data. DOT HS 808 679. Washington, D.C.: NHTSA.

Kahane, C.J. 1997. Relationships Between Vehicle Size and Fatality Risk in Model Year 1985–93 Passenger Cars and Light Trucks. NHTSA Technical Report, DOT HS 808 570. Springfield, Va: National Technical Information Services.

Klein, T., E. Hertz, and S. Borener. 1991. A Collection of Recent Analyses of Vehicle Weight and Safety. Technical Report DOT HS 807 677. Washington, D.C.: NHTSA.

Lund, A.K., and J.F. Chapline. 1999. Potential Strategies for Improving Crash Compatibility in the U.S. Vehicle Fleet (SAE 1999-01-0066). Vehicle Aggressivity and Compatibility in Automotive Crashes (SP-1442), 33–42. Warrendale, Pa.: SAE.

NRC (National Research Council). 1984. 55: A Decade of Experience. Special Report 204. Washington, D.C.: Transportation Research Board.

NRC. 1992. Automotive Fuel Economy: How Far Should We Go? Washington, D.C.: National Academy Press.

NRC-TRB (National Research Council, Transportation Research Board). 1996. Committee to Review Federal Estimates of the Relationship of Vehicle Weight to Fatality and Injury Risk. Letter report to Dr. Ricardo Martinez, Administrator, National Highway Traffic Safety Administration, June 12. Washington, D.C.: NRC.

O'Neill, B. 1998. The Physics of Car Crashes and the Role of Vehicle Size and Weight in Occupant Protection. Physical Medicine and Rehabilitation: State of the Art Reviews. Philadelphia, Pa.: Hanley & Belfus, Inc.

Partyka, S.C. 1995. Impacts with Yielding Fixed Objects by Vehicle Weight. NHTSA Technical Report, DOT HS 808-574, June. Washington, D.C.: NHTSA.

Partyka, S.C., and W.A. Boehly. 1989. Passenger Car Weight and Injury Severity in Single Vehicle Nonrollover Crashes. Proceedings of the Twelfth International Technical Conference on Experimental Safety Vehicles. Washington, D.C.: NHTSA.

Patterson, P. 1999. Reducing Greenhouse Gases in the U.S. Transportation Sector. SAE, Government and Industry Meeting, April 28, Washington, D.C.

Ross, M., and T. Wenzel. 2001. Losing Weight to Save Lives. A Review of the Role of Automobile Weight and Size in Traffic Fatalities, Report submitted to the committee on March 13, 2001.

Wang, M. 1996. GREET 1.0—Transportation Fuel Cycle Model: Methodology and Use, ANL/ESD-33. Argonne, Ill.: Argonne National Laboratory.

Wang, M., and H.S. Huang. 1999. A Full Fuel-Cycle Analysis of Energy and Emissions Impacts of Transportation Fuels Produced from Natural Gas. ANL/ESD-40. Argonne, Ill.: Argonne National Laboratory.

Weiss, M. A., J. B. Heywood, E. M. Drake, A. Schafer, and F. F. AuYeung. 2000. On the Road in 2020: A Life-Cycle Analysis of New Automobile Technologies. Energy Laboratory Report # MIT EL 00-003, October. Cambridge, Mass.: Massachusetts Institute of Technology (MIT).

Wood, D.P. 1997. "Safety and the Car Size Effect: A Fundamental Explanation." *Accident Analysis and Prevention* 29 (2): 139–151.

Attachment 4A

Life-Cycle Analysis of Automobile Technologies

Assessments of new automobile technologies that have the potential to function with higher fuel economies and lower emissions of greenhouse gases (GHGs) have been made by the Energy Laboratory at the Massachusetts Institute of Technology (MIT) (Weiss et al., 2000) and by the General Motors Corporation (GM), Argonne National Laboratory, BP, ExxonMobil, and Shell (General Motors et al., 2001, draft). Both studies compared fuels and engines on a total systems basis, that is, on a well-to-wheels (WTW) basis. These assessments provide an indication of areas of promising vehicle and fuel technology and benchmarks for likely increases in fuel economy and reduction of GHG emissions from the light-duty fleet over the next two decades. This attachment provides additional information on the emerging technologies and GHG emissions described in Chapters 3 and 4.

MIT's analysis was confined to midsize cars with consumer characteristics comparable to a 1996 reference car such as the Toyota Camry. It was assumed that, aided by the introduction of low-sulfur fuels, all technologies would be able to reduce emissions of air pollutants to levels at or below federal Tier 2 requirements. Only those fuel and vehicle technologies that could be developed and commercialized by 2020 in economically significant quantities were evaluated. General Motors et al. (2001) focused on the energy use of advanced conventional and unconventional power-train systems that could be expected to be implemented in the 2005 to 2010 time frame in a Chevrolet Silverado full-size pickup truck. The technologies were assessed on the basis of their potential for improving fuel economy while maintaining the vehicle performance demanded by North American consumers. Vehicle architectures and fuels analyzed in both studies are listed in Table 4A-1.

In the MIT study, the vehicle lifetimes and driving distances were assumed to be similar for all vehicles. The more advanced technologies were compared to an "evolved baseline" vehicle—a midsize passenger car comparable in consumer characteristics to the 1996 reference car, in which fuel consumption and GHG emissions had been reduced by about a third by 2020 through continuing evolutionary improvements in the traditional technologies currently being used. Figure 4A-1 summarizes energy use, GHG emissions, and costs for all the new 2020 technologies relative to the 1996 reference car and the evolved 2020 baseline car. (The battery-electric car shown is an exception in that it is not "comparable" to the other vehicles; its range is about one-

TABLE 4A-1 Vehicle Architecture and Fuels Used in the MIT and General Motors et al. Studies

MIT (Weiss et al., 2000)	General Motors et al. (2001)
1996 reference internal combustion engine (ICE)	Conventional (CONV) with spark ignition (SI) gasoline engine (baseline) [GASO SI CONV]
Baseline evolved ICE	CONV with SI E85 (85% ethanol and 15% gasoline by volume) engine [HE85 SI CONV]
Advanced gasoline ICE	CONV with compression-ignition direct-injection (CIDI) diesel engine [DIESEL CIDI CONV]
Gasoline ICE hybrid vehicle (HEV)	Charge-sustaining (CS) parallel hybrid electric vehicle (HEV) with SI E85 engine [HE 85 SI HEV]
Diesel ICE hybrid	CS HEV with CIDI diesel [DIESEL CIDI HEV]
CNG ICE hybrid	Gasoline fuel processor (FP) fuel cell vehicle (FCV) [GASO FP FC FCV]
Gasoline fuel cell (FC) hybrid vehicle	Gasoline (naphtha) FP fuel cell (FC) HEV [NAP FP FC HEV]
Methanol FC hybrid	Gaseous hydrogen (GH_2) refueling station (RS) FC HEV [GH_2 RS FC HEV]
Hydrogen FC hybrid	Methanol (MeOH) FP FCV [MEOH FP HEV]
Battery electric	Ethanol FP FC HEV [HEV 100 FP FC HEV]

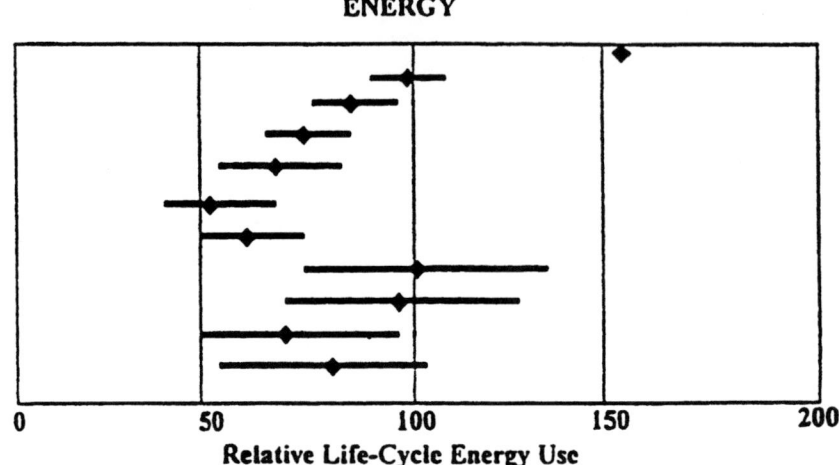
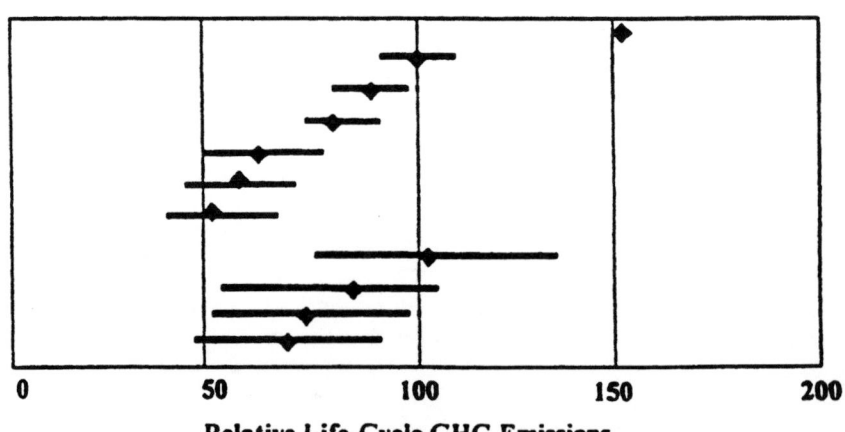
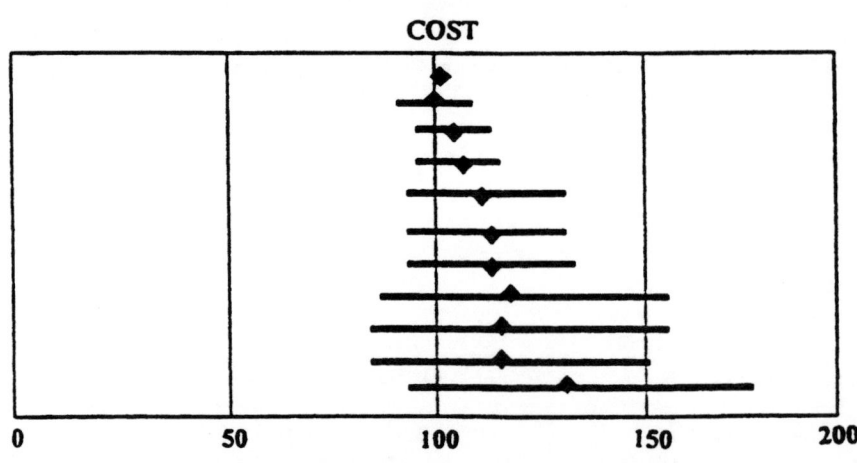

FIGURE 4A-1 Life-cycle comparisons of technologies for midsize passenger vehicles. NOTE: All cars are 2020 technology except for the 1996 reference car. On the scale, 100 = 2020 evolutionary baseline gasoline ICE car. Bars show estimated uncertainties. SOURCE: Weiss et al. (2000).

third less than that of the other vehicles.) The bars suggest the range of uncertainty surrounding the results. The uncertainty is estimated to be about ±30 percent for fuel-cell and battery vehicles, ±20 percent for hybrid electric vehicles (HEVs) using internal combustion engines (ICE), and ±10 percent for all other vehicle technologies.

MIT concludes that continued evolution of the traditional gasoline car technology could result in 2020 vehicles that reduce energy consumption and GHG emissions by about one-third relative to comparable vehicles of today at a cost increment of roughly 5 percent. More advanced technologies for propulsion systems and other vehicle components could yield additional reductions in life-cycle GHG emissions (up to 50 percent lower than those of the evolved baseline vehicle) at increased purchase and use costs (up to about 20 percent greater than those of the evolved baseline vehicle). Vehicles with HEV propulsion systems using either ICE or fuel-cell power plants are the most efficient and lowest-emitting technologies assessed. In general, ICE HEVs appear to have advantages over fuel-cell HEVs with respect to life-cycle GHG emissions, energy efficiency, and vehicle costs, but the differences are within the uncertainties of MIT's results and depend on the source of fuel energy. If automobile systems with drastically lower GHG emissions are required in the very long run future (perhaps in 30 to 50 years or more), hydrogen and electrical energy are the only identified options for "fuels," but only if both are produced from nonfossil sources of primary energy (such as nuclear or solar) or from fossil primary energy with carbon sequestration.

The results from the General Motors et al. study (2001) are shown in Figures 4A-2 and 4A-3. The diesel compression-ignition direct-injection (CIDI)/HEV, gasoline and naphtha fuel-processor fuel-cell HEVs, as well as the two hydrogen fuel-cell HEVs (represented only by the gaseous hydrogen refueling station and fuel-cell HEV in Figure 4A-2) are the least energy-consuming pathways. All of the crude-oil-based selected pathways have well-to-tank (WTT) energy loss shares of roughly 25 percent or less. A significant fraction of the WTT energy use of ethanol is renewable. The ethanol-fueled vehicles yield the lowest GHG emissions per mile. The CIDI HEV offers a significant reduction of GHG emissions (27 percent) relative to the conventional gasoline spark-ignited (SI) vehicle.

Considering both total energy use and GHG emissions, the key findings by General Motors et al. (2001) are these:

- Of all the crude oil and natural gas pathways studied, the diesel CIDI hybrid electric vehicle (HEV), the gasoline and naphtha HEVs, and the gaseous hydrogen fuel-cell HEVs were nearly identical and best in terms of total system energy use. Of these technologies, GHG emissions were expected to be lowest for the gaseous hydrogen fuel-cell HEV and highest for the diesel CIDI HEV.
- The gasoline-spark-ignited HEV and the diesel CIDI

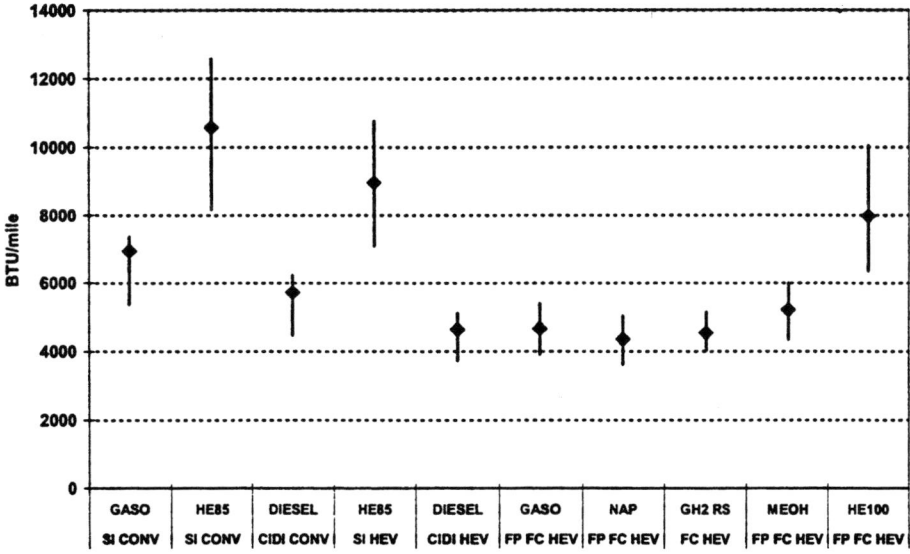

FIGURE 4A-2 Well-to-wheels total system energy use for selected fuel/vehicle pathways. SOURCE: General Motors et al. (2001).

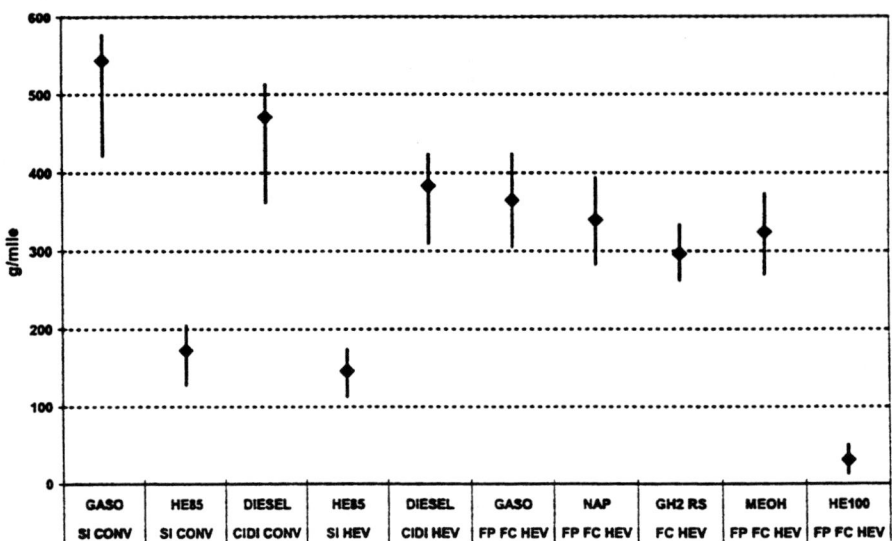

FIGURE 4A-3 Well-to-wheels greenhouse gas emissions for selected fuel/vehicle pathways. SOURCE: General Motors et al. (2001).

HEV, as well as the conventional CIDI diesel, offer significant total system energy use and GHG emission benefits compared with the conventional gasoline engine.
- The methanol fuel-processor fuel-cell HEV offers no significant energy use or emissions reduction advantages over the crude-oil-based or other natural-gas-based fuel cell HEV pathways.
- Bioethanol-based fuel/vehicle pathways have by far the lowest GHG emissions of the pathways studied.
- Major technology breakthroughs are required for both the fuel and the vehicle for the ethanol fuel-processor fuel-cell HEV pathway to reach commercialization.
- On a total system basis, the energy use and GHG emissions of compressed natural gas and gasoline spark-ignited conventional pathways are nearly identical. (The compressed natural gas [CNG] pathway is not shown in Figures 4A-2 or 4A-3).
- The crude-oil-based diesel vehicle pathways offer slightly lower system GHG emissions and considerably better total system energy use than the natural-gas-based Fischer-Tropsch diesel fuel pathways. Criteria pollutants were not considered.
- Liquid hydrogen produced in central plants, Fischer-Tropsch naphtha, and electrolysis-based hydrogen fuel-cell HEVs have slightly higher total system energy use and the same or higher levels of GHG emissions than gasoline and crude naphtha fuel-processor fuel-cell HEVs and electrolytically generated hydrogen fuel-cell HEVs.

References

General Motors Corporation, Argonne National Laboratory, BP, Exxon Mobil and Shell. 2001. Wheel-to-Well Energy Use and Greenhouse Gas Emissions of Advanced Fuel/Vehicle Systems. North American Analysis, April. Final draft.

Weiss, Malcolm A., John B. Heywood, Elisabeth M. Drake, Andreas Schafer, and Felix F. AuYeung. 2000. On the Road in 2020: A Life-Cycle Analysis of New Automobile Technologies. Energy Laboratory Report # MIT EL 00-003, October. Cambridge, Mass.: Massachusetts Institute of Technology.

5

Potential Modifications of and Alternatives to CAFE

WHY GOVERNMENTAL INTERVENTION?

Why should the government intervene in the fuel economy decisions of consumers and manufacturers? This section discusses the underlying rationales.

Environmental and International Oil Market Impacts

Fuel economy decisions can be distorted if the market price of gasoline—the price that motivates decisions—fails to take account of the environmental impacts of gasoline use, the impacts of oil consumption on world oil prices, or the impacts of oil consumption on vulnerability to oil market disruptions. And, absent intervention, the resulting distortions would result in a fleet of new vehicles with fuel economy lower than what is optimal for the United States as a whole. Appropriately designed and scaled interventions can successfully mitigate these distortions and thereby enhance overall welfare. This chapter examines the appropriate scale of interventions and explores alternative policy instruments that could reasonably be expected to enhance overall welfare.

The primary environmental issue is the emissions of carbon dioxide (CO_2), a normal combustion product of hydrocarbons, and the resultant impacts of atmospheric accumulations of CO_2 on global climate change.[1] The amount of CO_2 released from driving is directly proportional to the amount of gasoline consumed. There is also an environmental cost associated with releases of hydrocarbons and toxic chemicals from the gasoline supply chain.

The second issue is the impact of increased oil consumption on the world oil market and on oil market vulnerability. The price of oil imported into the United States exceeds the competitive level because of the Organization of Petroleum Exporting Countries' (OPEC's) market power; greater U.S. oil consumption could further increase the import price. In addition, international oil market disruptions could lead to economic losses in the United States. The greater the consumption of petroleum products, the more vulnerable the United States is to such disruptions.

These factors together imply that there are costs of increasing gasoline use in addition to those seen by the individual consumer. These additional costs are referred to collectively as externalities (external costs).

Since the rationale for fuel-use-reducing market interventions is the existence of external costs, the magnitude of these external costs determines the appropriate scale, or strength, of the interventions. Economic efficiency requires that consumers face the full social cost (including the external cost) associated with gasoline use, or be induced to act as if they faced those full costs. Therefore, quantification of these external costs is important for policy analysis. A later section discusses this quantification.

Unresolved Issues of Governmental Intervention

Some analysts argue that, even in the absence of any environmental and international oil market impacts, the United States should intervene in automobile markets to require higher fuel economy than would be chosen by manufacturers and consumers absent market intervention. These analysts argue that there are reasons to believe that the market choices for fuel economy are not efficient, even absent these externalities.[2]

The net value of major increases in fuel economy is, at most, some hundreds of dollars to new car buyers even if

---

[1] Vehicles also emit criteria pollutants, but these pollutants are regulated on a grams-per-mile basis, with allowable emissions not dependent on automotive fuel economy. Therefore, varying the fuel economy of new vehicles is unlikely to cause significant variations in emissions of criteria pollutants.

[2] An intermediate position, held by at least one committee member, is that there should be no intervention absent environmental and international oil market impacts, but that when one combines environmental externalities and oil market problems with the imperfections in the markets for fuel economy, the case for action becomes strong.

fuel savings over the entire life of the vehicle are considered. This is because the net value is the difference between the discounted present value of fuel savings and the cost to the consumers of added technologies to achieve these savings. But if buyers consider the fuel costs only over the first few years they intend to own the vehicle, the perceived net value from these costly changes could be very low—even negative—and consumers might prefer to not pay for such changes.

In order to implement significant fuel economy increases, manufacturers must completely redesign all the vehicles they make. Manufacturers would decide whether to make very expensive and risky investment decisions, in the expectation of a small, uncertain advantage. According to this view, there is a lot of inertia in the market choices determining fuel economy, and one cannot be sure that manufacturers and consumers would ever arrive at the optimal fuel economy level.

In addition, although there is good information available to consumers on the fuel economy of new vehicles, the information is not perfect. And, since consumers do not know what the future price of fuel will be, they may underestimate or overestimate future fuel costs. Consumers typically do not actually compute the discounted present value of fuel savings before buying a car. Most importantly, there is no pure "price" of higher fuel economy facing car buyers; instead, they must infer how much greater fuel economy will cost by comparing different vehicles. Thus, these consumers may buy vehicles with fuel economy that is higher or lower than what they would have chosen had perfect information been available.

Finally, from this point of view, although automobile companies compete intensely with each other, the automobile market is not perfectly competitive—in fact, it is more adequately described as oligopolistic. In oligopolistic markets, companies may choose levels of fuel economy that are higher or lower than they would have chosen if the markets were perfectly competitive. For all of these reasons, this viewpoint maintains that there is no guarantee that markets will achieve economically optimal levels of fuel economy; rather, the levels could be either too high or too low.

Committee members differ in their beliefs about the quantitative importance of these issues and whether they justify government intervention to regulate fuel economy.

Those supporting the viewpoint described here suggest that because automobile manufacturers might systematically produce vehicles with lower-than-optimal fuel economy, the government should intervene in the markets and should require manufacturers to increase fuel economy to some "correct" level. This viewpoint requires calculation, external to the automobile manufacturers, of the correct level, at least for each type of vehicle. That correct level of fuel economy might be taken to be the cost-efficient level—that is, the level at which the estimated cost of additional fuel economy improvements would be just equal to the estimated discounted present value of additional fuel cost savings over the entire life of the vehicle, using some estimate of future gasoline prices and some specified discount rate for future fuel cost savings. These fuel economy levels might correspond to the 14-year case described in Chapter 4, if all assumptions underlying that calculation turned out to be accurate.

Those rejecting the viewpoint argue that it is in the interests of automobile manufacturers to estimate the preferences of their customers and others they wish to attract as their customers. Therefore, manufacturers provide levels of fuel economy that, in their estimation, best reflect the trade-offs potential customers would make themselves.

But manufacturers realize, too, that vehicle buyers differ greatly from one another, including in the trade-offs they are willing to make between vehicle purchase price and fuel economy. Accordingly, the various manufacturers offer different makes and models for sale, with a range of fuel economies. Potential customers are free to choose vehicles that correspond to their particular preferences. Some will wish to purchase vehicles with fuel economy corresponding to the 14-year case. Others value vehicle purchase price and may prefer vehicles that use more gasoline but are less expensive. For them, the best choice might be a vehicle with fuel economy corresponding to the 3-year case from Chapter 4.

From this perspective, if the government requires fuel economy to correspond to the 14-year case, then those people who prefer to purchase vehicles corresponding to the 3-year case would be harmed. They would have to pay more money to purchase a vehicle. Although they would subsequently spend less on gasoline, the gasoline savings would not be sufficient to compensate them for the increased vehicle purchase price. Conversely, if the government were to require fuel economy to correspond to the 3-year case, then those who would prefer to purchase vehicles corresponding to the 14-year case would be harmed. Although they would save money on new vehicle purchases, the savings would not be sufficient to compensate them for the additional price they would pay for gasoline over the life of the vehicle. This perspective notes that absent government intervention, each type of consumer can be satisfied because the competing manufacturers will offer a range of options from which consumers can select. If fuel economy is regulated, the range of consumer choice may be sharply diminished, and some people will be harmed.

This debate has not been resolved within the committee, nor within the community of policy analysts. However, as is clear from Tables 4-2 and 4-3 in Chapter 4, the difference between the cost-efficient fuel economies of vehicles in the 14-year case and the 3-year case is large. Thus, if most consumers had preferences corresponding to the 3-year case, yet fuel economy standards were set to correspond to the 14-year case, most consumers would be made economically worse off by such governmental regulation. Conversely, if most consumers had preferences corresponding to the 14-year case, yet automobile manufacturers offered only vehicles corresponding to the 3-year case, most consumers would be made economically worse off by the government's failure to intervene to increase fuel economy.

Other Issues Not Considered

The committee has explicitly not relied on rationales other than environmental and international oil market impacts. It is sometimes asserted that increases in fuel economy would reduce tailpipe emissions of criteria pollutants such as NO_x or volatile organic compounds. But these criteria pollutants are regulated independent of fuel economy, so that each vehicle, when new, must have emissions below a federally mandated (or a tighter state-mandated) number of grams per mile. Since allowable grams per mile can be expected to remain the binding constraint and since allowable grams per mile does not depend on fuel economy, criteria pollutants do not provide a rationale for intervening to increase fuel economy. However, some research suggests that once the control systems of vehicles deteriorate, there is a relationship between fuel economy and emissions. And some particularly high-fuel-economy vehicles now operate well below the grams-per-mile standards, although this is less likely to occur with higher fuel economy standards and/or with tightened standards for criteria pollutant emissions. The trend toward increasing the required warranty times for pollution control systems is likely to render this phenomenon moot for the purpose of evaluating fuel economy standards.

Second, some critics believe manufacturers overestimate costs, that it would cost little to improve fuel economy, and that there could even be manufacturing cost savings associated with such improvements. The cost estimates presented by the committee in Chapter 3 are, as might be expected, lower than some from industry. Overall, however, the committee concludes that improving fuel economy significantly will raise the price of vehicles significantly.

Third, some critics believe that consumers who care more about performance characteristics, such as acceleration, than about fuel economy are irrational. But this represents a difference in tastes—the consumers and the critics value different things—not a difference in rationality.

Finally, other industry observers have estimated external costs of driving in addition to the costs identified above, such as the costs of road congestion and policing. However, such costs will be unaffected by fuel economy and therefore do not provide a reason for market intervention to improve fuel economy.

Quantifying Environmental and International Oil Market Costs

One product of the combustion of hydrocarbon fuels such as gasoline or diesel in internal combustion engines is CO_2. Scientific discussions of greenhouse gases typically refer to CO_2 emissions in terms of the weight of the carbon (C) contained in the CO_2, and it is this terminology that the committee uses in this report. The combustion of each gallon of gasoline releases 8.9 kg of CO_2, or 2.42 kg of C in the form of CO_2.[3] The environmental and economic consequences of these releases are not included in the price of gasoline and are part of the environmental externalities of gasoline use.

To quantify the environmental externalities associated with such CO_2 releases, in principle one could directly measure the various consequences of additional CO_2 releases and place monetary values on each consequence. Although estimates have been made of the costs to agriculture, forestry, and other economic activities, estimating the marginal costs of environmental degradation, species extinction, increased intensity of tropical storms, and other impacts beyond commercial activities has proven highly controversial. A wide range of estimates appears in the literature, from negative values to values well over $100 per metric ton (tonne) of C. Public debate suggests that many people would estimate values even outside of the published range of estimates, particularly because there are many possible, although highly unlikely, events that could be very harmful or very beneficial. The committee has used a figure of $50/tonne C as an estimate of the environmental externality of additional carbon emissions, although this figure is significantly higher than typical estimates in the published literature.[4] This estimate translates into a cost of $0.12/gal (gasoline), the value used in the examples in this chapter. A range of cost estimates from $3/tonne to $100/tonne would give a range of estimated external costs from $0.007/gal to $0.24/gal of gasoline.

A second environmental cost of gasoline use is related to the hydrocarbon and toxic chemical releases from the gasoline supply chain, including oil exploration and recovery, oil refining, and distribution (tanker, pipeline, or tanker truck distribution, and gasoline retail sales). The more gasoline used, the greater will be the amount of hydrocarbons and toxics released. However, the supply chain is tightly regulated, and releases per gallon of gasoline now are very limited. Marginal costs of this environmental impact are small.

---

[3]Combustion of gasoline releases about 19.36 kg of C per million Btu (MMBtu) of gasoline. A barrel of gasoline has an energy content of about 5.25 MMBtu; there are 42 gallons per barrel.

[4]At one of the committee's public meetings, a representative of an environmental advocacy organization indicated that there was much uncertainty but offered a figure of $50/tonne. That figure is viewed as high, but not implausibly high, by committee members who have been involved in the global climate change debates.

[5]This issue was examined in Delucchi et al. (1994). That study showed various estimates for the value of hydrocarbon reductions from fuel economy improvements, depending on assumptions about the value of reducing hydrocarbon emissions, upstream control effectiveness, possible benefits of refueling, and evaporative losses. Based on this work, $0.02/gal seemed a reasonable estimate, given current emissions control trends and using $1,000/ton (1990 dollars) marginal damage of hydrocarbon emissions.

The committee used an estimate of a $0.02/gal for the total of these external costs in its calculations.[5]

The Organization of Petroleum Exporting Countries (OPEC) operates as a cartel that restricts the supply of oil to escalate its price above the free-market level. The greater the consumption of oil, the higher will be its price. Since the higher price would apply to all oil imports, not just to the increased consumption, the financial cost to the United States of increased oil use exceeds the market payment for the increased amount. The additional financial cost of importing more oil, often referred to as the monopsony component of the oil import premium, was much studied after the energy crises of 1973–1974 and 1979–1980.

The volatility of oil prices also creates problems. Past oil shocks may have caused significant macroeconomic losses to the U.S. economy. As U.S. oil consumption rises, so does the vulnerability to such disruptions.[6] The additional vulnerability cost, often referred to as the security component of the oil import premium, was also much studied in the wake of past energy crises.

Observations of the world oil market since the 1980s suggest that the monopsony component of the oil import premium is small, primarily because the impact of U.S. oil consumption on world prices has proven to be smaller than once thought. These observations also suggest that the security premium also has become smaller, because the U.S. economy is now less vulnerable to oil price volatility than in the 1970s and the early 1980s: the United States has become significantly more energy efficient, its expenditure on oil relative to the gross domestic product (GDP) has declined, there now are more mechanisms for cushioning oil shocks, and the nation's ability to manage the overall economy has improved greatly. Therefore, the marginal cost to the United States of oil consumption is now considerably smaller than previously estimated and is likely to remain so. However, this issue may become more important if the world price of oil rises. In addition, some analysts would argue that the concentration of oil use for transportation might be relevant, since there are few substitutes for oil in this sector.

For its examples, the committee used an estimated external marginal cost of oil consumption of $5.00/bbl of oil for the combined monopsony component and security component of the oil import premium, although the cost could be smaller or larger than this figure.[7] This estimate translates into a cost of $0.12/gal of gasoline. An oil import premium range from $1/bbl to $10/bbl would give a range of estimated external costs from $0.02/gal to $0.24/gal of gasoline.

It should be emphasized that the monopsony component of the oil import premium is the marginal cost of increasing oil use. It includes neither the entire benefit to the United States of "solving" the problem of noncompetitive pricing by the OPEC nations nor the entire benefit of increasing international stability in world oil markets (or, equivalently, the cost of not solving these problems). These problems cannot be solved completely by changing the amount of oil consumed in the United States.

Combining the $0.12/gal marginal cost estimate for CO_2 externalities, the $0.12/gal for international oil external marginal costs, and the $0.02/gal figure for externalities in the gasoline supply chain, the committee uses a total external marginal cost of additional gasoline use of $0.26/gal in all of the examples, although estimates as high as $0.50/gal or as low as $0.05/gal are not implausible and estimates well outside of that range cannot be rejected out of hand.

ALTERNATIVE POLICIES—SUMMARY DESCRIPTION

Presentations to the committee, a review of published literature, and committee deliberations identified many possible modifications to the current CAFE system as well as other approaches to improving fuel economy. The various approaches are generally not mutually exclusive but normally can be used alone or in combination with others.

These changes can generally be grouped into four broad classes:

- *Retain the basic CAFE structure.* This approach would keep CAFE basically intact but would modify some elements that are particularly troublesome.
- *Restructure fuel economy regulations.* This approach would restructure CAFE with alternative regulatory or incentive policies directed at the fuel economy of new vehicles.
- *Adopt energy demand-reduction policies.* This broader approach is designed to reduce either gasoline consumption or consumption of all fossil fuels.
- *Pursue cooperative government/industry technology strategies.* This approach would attempt to advance automotive technologies to greatly improve fuel economy.

Retain the Basic CAFE Structure

This class of policies would keep CAFE basically intact but would modify some troublesome elements, particularly those involving domestic versus import production and the definitions used to classify vehicles as trucks or passenger cars.

---

[6]Some people believe that military expenditures will also increase as a function of gasoline price (above and beyond mere fuel costs). However, the committee has seen no evidence to support that belief.

[7]Work by Leiby et al. (1997) provides estimates of $3.00/bbl. The Energy Modeling Forum (1982) study *World Oil*, using nine different mathematical models, estimated that in 2000, oil price would be increased by between 0.8 percent and 2.9 percent for every million barrels per day of oil import reduction. Applying that same percentage to current prices would give a monopsony component between $2.20 and $8.20/bbl. The vulnerability component was much smaller in that study.

The Two-Fleet Rule Differentiating Between Domestic and Imported Cars

Currently, each manufacturer must meet the CAFE standard separately for its domestically produced fleet of new passenger cars and for its imported fleet. Averaging[8] is allowed within each of these two fleets but is not allowed across the two fleets. A domestically produced fleet that significantly exceeded the CAFE standard could not be used to compensate for an imported fleet that failed to meet the standard and vice versa. This requirement that the domestically produced fleet and the imported fleet each separately meet the CAFE standard is referred to as the two-fleet rule. No such requirement exists for light trucks. This distinction could be removed and the fuel economy standard could apply to the entire new car fleet of each manufacturer.

Classification of Vehicles As Trucks vs. Passenger Cars

The distinction between passenger cars and light-duty trucks could be redefined to correspond more closely to the original distinction between the two classes—passenger vehicles and work/cargo vehicles. Alternatively, the incentives for manufacturers to classify vehicles as trucks could be reduced or eliminated. The provisions for flexible-fuel vehicles could be eliminated or redefined to ensure they will apply only to vehicles that will often use alternative fuels.

Restructure Fuel Economy Regulations

These policies more fundamentally restructure CAFE with alternative regulatory or incentive policies directed at fuel economy of new vehicles.

Tradable Fuel Economy Credits

There would be an increase in the economic efficiency and the flexibility of the CAFE system if a market-based system of tradable fuel economy credits were created, under which automobile manufacturers could sell fuel economy credits to other manufacturers and could buy credits from other manufacturers or from the government. This system would be similar in many respects to the successful system now used for trading sulfur emission credits among electricity power plants.

Feebates

Feebates is an incentive mechanism that uses explicit government-defined fees and rebates. Vehicles with a fuel economy lower than some fuel economy target pay a tax, while vehicles with a fuel economy higher than the target receive a rebate. Such systems could be designed to be revenue neutral: the tax revenues and the subsidies would just balance one another if the forecasted sales-weighted average fuel economy (or average per-mile fuel consumption) turned out as predicted.

Attribute-Based Fuel Economy Targets

The government could change the way that fuel economy targets[9] for individual vehicles are assigned. The current CAFE system sets one target for all passenger cars (27.5 mpg) and one target for all light-duty trucks (20.7 mpg). Each manufacturer must meet a sales-weighted average (more precisely, a harmonic mean—see footnote 8) of these targets. However, targets could vary among passenger cars and among trucks, based on some attribute of these vehicles—such as weight, size, or load-carrying capacity. In that case a particular manufacturer's average target for passenger cars or for trucks would depend upon the fractions of vehicles it sold with particular levels of these attributes. For example, if weight were the criterion, a manufacturer that sells mostly light vehicles would have to achieve higher average fuel economy than would a manufacturer that sells mostly heavy vehicles.[10]

Uniform Percentage Increases

There have been proposals that would require each manufacturer to improve its own CAFE average by some uniform percentage, rather than applying the targets uniformly to all manufacturers. This is often referred to as the uniform percentage increase (UPI) standard.

Adopt Energy Demand-Reduction Policies

There are several alternatives aimed more broadly at reducing motor fuel consumption or all fossil fuel consumption. They could be part of a more comprehensive energy policy. If these more broadly based alternatives were implemented, they could be used in place of or along with the instruments aimed directly at new vehicle fuel economy.

---

[8]Under CAFE, the "average" fuel economy is the sales-weighted harmonic mean of fuel economies of the individual vehicles sold by the manufacturer. Mathematically, a standard on the sales-weighted harmonic mean of fuel economies of individual vehicles is exactly equivalent to a standard on the sales-weighted average of per-mile fuel consumption of individual vehicles. In this chapter, the word "average" or "averaging" is used to denote this harmonic mean of fuel economies or average of per-mile consumption.

[9]Throughout this chapter, the word "target" is applied to the goal for fuel economy, or per-mile fuel consumption, of individual vehicles or groups of vehicles. The word "standard" is used to denote a regulatory rule that must be met. Under the current CAFE system, regulations do not require that each car or truck meet any particular target, although CAFE requires the corporation to meet the standards for the aggregate of all passenger cars and the aggregate of all trucks.

[10]Targets could also be normalized, for example, by expressing them in terms of weight-specific fuel consumption—for example, gallons used per ton of vehicle weight per 100 miles.

Gasoline Tax

The current federal excise tax on gasoline could be increased. A tax increase would provide direct incentives for consumers to buy and for manufacturers to produce higher-fuel-economy new vehicles and would also provide incentives to reduce the use of all new and existing vehicles.

Carbon Taxes/Carbon Cap-and-Trade Systems

In order to address problems of global greenhouse gas release, the United States could impose a carbon tax or could adopt a carbon cap-and-trade system. Under these systems, the total annual emissions of carbon dioxide would be limited or capped, rights to emit carbon would be allocated or auctioned off, and these rights would then be tradable among firms. In either system, the price of energy would increase, on a fuel-by-fuel basis, roughly in proportion to the amount of CO_2 released from combustion of that fuel. Such plans would provide a broad-based incentive to use less of all fossil fuels, especially those that are particularly carbon-intensive.

Pursue Cooperative Government/Industry Technology Strategies

The final class of strategies would attempt to create dramatic changes in automotive technologies, changes that could greatly alter the economy of fuel consumption or the types of fuels used.

Partnership for a New Generation of Vehicles (PNGV)

A particular, ongoing example of such strategies is the Partnership for a New Generation of Vehicles (PNGV) program, a private-public research partnership that conducts precompetitive research on new vehicle technologies. One of its goals is to create marketable passenger cars with fuel economy up to 80 mpg.

Technology Incentives

The government could provide tax or other incentives to manufacturers or consumers for vehicles that embody new high-efficiency technology. Such incentives would encourage manufacturers to pursue advanced technology research and to bring those new technologies to market and would encourage consumers to purchase vehicles that use them.

MORE COMPLETE DESCRIPTIONS OF THE ALTERNATIVES

Retain the Basic CAFE Structure

Classification of Vehicles

When CAFE regulations were originally formulated, different standards were set for passenger vehicles and for work/cargo vehicles. Work/cargo vehicles (light-duty trucks that weigh less than 8,500 lb gross vehicle weight) were allowed higher fuel consumption because they needed extra power, different gearing, and less aerodynamic body configurations to carry out their utilitarian, load-carrying functions. At that point, light-truck sales accounted for about 20 percent of the new vehicle market. However, as one observer noted, "the 1970s working definition distinction between a car for personal use and a truck for work use/cargo transport, has broken down, initially with minivans, and more recently with sport utility vehicles and other 'cross-over' vehicles that may be designed for peak use but which are actually used almost exclusively for personal transport."[11]

The car/truck distinction bears critically on fuel economy considerations. Trucks are allowed to meet a lower CAFE standard, 20.7 mpg versus 27.5 mpg for cars, and their market share has increased enormously. Vehicles classed as light-duty trucks now account for about half the total new vehicle market. The car/truck distinction has been stretched well beyond its original purpose. For example, the PT Cruiser, a small SUV that can carry only four passengers and cannot tow a trailer, is considered a truck, while a large sedan that can carry six passengers while towing a trailer is considered a car.

Two kinds of change might alleviate these problems:

- Redefine the criteria determining whether a vehicle is classified as a car or a truck or
- Sharply reduce the economic incentives for manufacturers to classify their vehicles as trucks.

Fuel economy regulators might tighten the definition of a truck. The Environmental Protection Agency (EPA) has already done so: for example, it classifies the PT Cruiser as a car for purposes of the emissions standards. EPA and the National Highway Traffic Safety Administration (NHTSA) have considerable regulatory discretion to implement such changes after a rulemaking process.

The economic incentives for manufacturers to classify their vehicles as trucks come from both the CAFE standards and the gas guzzler tax. Because CAFE standards require much greater fuel economy for cars than for trucks and because they impose a binding constraint on manufacturers, CAFE standards create a strong incentive to make design changes in vehicles that allow them to be classified as trucks, whenever such changes are possible at modest cost. (For example, the PT Cruiser was designed with removable rear seats, which allows it to be classified as a truck.)

Reductions in the differential between CAFE standards for trucks and standards for cars would therefore reduce these economic incentives. At one extreme, eliminating the differ-

---

[11] J. Alston, EPA, Letter to the committee dated April 16, 2001.

ential between truck and car CAFE standards would eliminate all CAFE-derived incentives for manufacturers to have vehicles classified as trucks.[12] This is the approach EPA will use for emission standards: beginning in 2009, it will treat light-duty trucks the same as it treats cars under the Tier 2 emission regulations.[13]

In addition, the gas guzzler tax, discussed in Chapter 2, imposes a large tax on passenger cars but not on trucks. For example, a passenger car with a fuel economy of 20 mpg would face a gas guzzler tax of $1,000, while a truck with similar fuel economy would face no tax. At 12 mpg, the tax on a car would be $7,000. This tax provides a large financial incentive to design any vehicle expected to have low fuel economy in such a way as to assure that it will be classified as a truck, so that reducing or eliminating the tax would likewise reduce the incentive to ensure that vehicles are classified as trucks.

The distinction between cars and trucks extends to the processes for determining fuel economy standards. CAFE standards for passenger cars are set legislatively, with a long time horizon. CAFE standards for trucks are set by rulemaking within NHTSA, with shorter time horizons. Integration of the processes, so that trucks and passenger cars are subject to equivalent processes and equivalent time horizons for regulatory decision making, might lead to more consistency of treatment among vehicle types.

A related problem involves flexible-fuel vehicles. In calculating new car fuel economy for CAFE compliance purposes, a flexible-fuel vehicle is currently deemed to have a fuel economy 1.74 times as high as its actual fuel economy, with a 1.2 mpg maximum total increase per manufacturer from this flexible-fuel vehicle adjustment. This adjustment is based on a legislative assumption that 50 percent of the fuel such vehicles use would, on average, be E85, including only 15 percent petroleum.[14]

However, few of these vehicles ever use any fuel other than gasoline. Estimates from the Energy Information Administration (EIA) suggest[15] that for 1999, there were 725,000 vehicles capable of using E85, but only 3.1 percent of them were using any E85 at all. Total E85 consumption in 1999 was 2 million gallons, or only 92 gallons for each of the 3.1 percent of the vehicles using some E85 (EIA, 2001). Therefore, it is likely that even these vehicles were using E85 for less than 25 percent of their fuel requirements. In total, less than 1 percent of the fuel used in these vehicles seems to be E85, and more than 99 percent seems to be gasoline. Thus, the current incentives to produce such vehicles lead to increased costs and lower fleet fuel economy without corresponding benefits.

Distinction Between Domestic and Imported Passenger Cars

Current CAFE regulations make a distinction between domestic and imported passenger cars. The distinction is based on the proportion of the car that is manufactured in the United States; an import is defined as a car with less than 75 percent domestically produced content. Imports and domestically produced vehicles constitute two separate car fleets under CAFE regulations. Under these regulations the domestic fleet and the import fleet of passenger cars must separately meet the same 27.5 mpg standard. There is no such rule for trucks.

The two-fleet rule was added to the original CAFE legislation to protect domestic employment. There was a concern that U.S. manufacturers might decide to import their small cars rather than continue to make them in the United States, with a resultant loss of jobs. At one time the two-fleet rule made it impossible for the Big Three to increase imports to help meet their domestic fleet CAFE obligations.

Over time, however, foreign manufacturers have moved production to the United States. Now the two-fleet rule can just as well provide incentives for manufacturers to reduce domestic production and increase import production to help meet CAFE obligations. In addition, under the North American Free Trade Agreement (NAFTA), manufacturers can count vehicles produced in Canada as part of their domestic fleet and will soon be able to count those produced in Mexico. Thus it appears to the committee that the two-fleet rule no longer serves to protect U.S. employment.

Presentations to the committee during its open meetings noted that the rule produces perverse results and increases costs. For example, representatives of American Honda Motors testified that Honda ships Accords from Japan to assure that Honda Accords are classified as an import under CAFE. Reporting on the testimony on March 19, 2001, in *Automotive News*, Stoffer (2001) wrote as follows:

> "Last year, by selling about 87,000 Japan-built Accords, to go with the 317,000 built in Ohio, Honda was able to keep the average domestic content of the whole model line below the cutoff point of 75 percent," Honda spokesman Art Garner said. That means the Accord remains classified an import model line and helps Honda keep the average fuel economy of its imported fleet comfortably above the 27.5 mpg standard. Without the more efficient Accord, the imported fleet would consist entirely of the more performance-oriented Acuras and the Honda Prelude.

A second example was discussed in the same *Automotive News* story: "In the early 1990s, Ford Motor Co. intentionally reduced domestic content of the Crown Victoria and Grand Marquis because of CAFE."

---

[12]This statement is not a recommendation that the standards for trucks be made identical to those for passenger cars, rather a simple observation about the incentives that are created by the differential nature of these standards.

[13]The bin structure of Tier 2 emissions standards provides more flexibility for manufacturers to meet the standards, but the requirements are identical for passenger cars and light-duty trucks.

[14]E85 consists of 85 percent ethanol and 15 percent gasoline.

[15]The data appear online at <http://www.eia.doe.gov/cneaf/alternate/page/datatables/atf1-13_00.html>.

The United Auto Workers (UAW) union stated during the committee's open meetings that it supports continuation of the two-fleet rule and indicated it believes the rule continues to protect American jobs. In response to UAW communications, the committee sought to identify research or analysis that tends to validate that position but could not find any. It is possible that the rule provides some protection for existing jobs, but it appears likely that removing it would have little industrywide impact. Since the two-fleet rule increases costs to consumers, the committee believes it is no longer justifiable and should be eliminated.

Restructure Fuel Economy Regulations

Tradable Fuel Economy Credits

The existing CAFE system already allows a manufacturer to accumulate CAFE credits if its fleet mix exceeds the standard. These credits may be carried forward and used to offset future CAFE deficits by the same manufacturer. The idea of tradable fuel economy credits (tradable credits, for short) carries this flexibility one step further: Manufacturers could also be allowed to sell and buy credits among themselves or to buy credits from the government.

Under this system, fuel economy targets would be set, either uniform targets as in the current CAFE system, or attribute-based targets, as discussed below. Each manufacturer would be required either to meet these targets or to acquire sufficient credits to make up the deficit. The credits could be purchased from the government or from other automobile manufacturers. A manufacturer whose new vehicle fleet had greater fuel economy than the overall target would acquire credits. These credits could be saved for anticipated later deficits or could be sold to other manufacturers. Credits would be equal to the difference between the projected gasoline use over the life of the vehicle, using a legislatively deemed total lifetime vehicle miles, and the projected lifetime gasoline use of vehicles just meeting the target.

As an example, assume that a uniform target for cars of 30 mpg is legislated and the vehicle lifetime is deemed to be 150,000 miles. This implies a lifetime target fuel consumption of 5,000 gal/vehicle (150,000 miles; 30 mpg). A manufacturer that sold 1 million cars with an average expected lifetime fuel consumption of 4,500 gal each (33.33 mpg) would acquire 500 credits per car for each of 1 million cars, or 500 million credits. A manufacturer that sold 1 million cars with an average expected lifetime fuel consumption of 5,500 gal each (27.27 mpg) would need to purchase 500 million credits.

The government would assure that prices for tradable credits would not exceed some ceiling price by offering to sell credits to any manufacturer at some predetermined offer price. The offer price could be set equal to the estimated external costs per gallon of gasoline use.[16] If external costs (e.g., greenhouse gas emissions and international oil market) are estimated to be $0.26/gal, the government would offer to sell credits at a price of $0.26 per 1-gallon credit.

The availability of credits from the government is important because it represents a safety valve preventing excessive costs to manufacturers (and consumers) in the event that unforeseen market changes or errors in setting targets make attaining the target more costly than originally projected.[17] The market-clearing price of tradable credits would never exceed the government offer price, because a buyer of credits could always turn to the government if the price of credits were above the government offer price.

Suppose the marginal cost of reducing gasoline use enough to meet the target was greater than the sum of the gasoline price plus the market price of credits: Manufacturers could buy credits without being forced to install overly expensive technology or to make changes to vehicle attributes that could damage sales. Conversely, if the marginal cost of reducing gasoline use to meet the fuel economy target was less than the sum of gasoline price plus market price of credits, the manufacturer would choose to make changes necessary to meet or to exceed the fuel economy target. Since the decisions would be made by and the resulting financial costs borne by the manufacturer, the manufacturer would have a motivation to correctly estimate the costs of fuel economy increases. Under this system, the manufacturer could respond to automotive market conditions but would still have an enhanced incentive to increase fuel economy.

In comparison with the current CAFE system, a tradable credits system would *increase* the range of options available to manufacturers. Currently, manufacturers have two options: They can meet the standards or they can pay the government a civil penalty.[18] Under a system of tradable fuel economy credits, manufacturers would have more options: They could meet the targets, they could pay the government for credits, or they could purchase credits from other manufacturers. They would be free to choose.

Similarly, in comparison with the current CAFE system, a tradable credits system would increase the range of attractive options available to manufacturers whose fuel economy exceeds the target. Under CAFE, such manufacturers have no incentives to further increase fuel economy. But under a tradable fuel economy credits system, they would have the option of further increasing fuel economy and receiving additional credits that they could sell to other manufacturers.

---

[16]Applying this rule and using a reasonably accurate estimate of the lifetime miles of vehicles is economically efficient as long as the external cost per gallon of gasoline use during the future vehicle life has the same value for current decision making as the estimate of external costs used for the regulation.

[17]In addition, the safety valve limits the exercise of market power in the market for tradable fuel economy credits. Such market power could otherwise become an important problem if only a very small number of manufacturers were selling tradable credits.

[18]The penalty is currently $5.50 for every 0.1 mpg by which the manufacturer misses the standard.

The tradable credits system would have another advantage, especially if the sales price of tradable credits were made public.[19] Debates about environmental standards usually involve disputes about implementation cost: Those favoring regulation contend that the standards will be cheap to implement, while manufacturers contend that the standards will be too expensive. The sales price of credits will reflect the marginal costs of fuel economy improvements, since manufacturers can be expected to increase fuel economy to the point at which the marginal cost of fuel savings equals the sum of gasoline price plus the market price of tradable credits.

It should be noted that a similar tradable credits scheme has been used for some time in the electrical power industry to reduce sulfur emissions.[20] There is general agreement that tradable credits have been highly successful: They have reduced the economic cost of compliance, and they have reached the achieved environmental goals.

Feebates

Like tradable fuel economy credits, a feebate system[21] is an incentive mechanism that can be used with almost any method of specifying fuel economy targets. Under a feebate system, target fuel economies would be set, either uniform targets or attribute-based targets. Fees would be imposed on new vehicles with mpg's lower than the target; the lower the mpg, the greater the fee. Rebates would be provided to manufacturers of new vehicles with mpg's above the target. These fees and rebates would be aggregated across all vehicles sold by a single manufacturer, which would receive, or make, a single payment.

Feebates could be designed to be revenue neutral, with fee revenues and rebates balancing each other. However, actual revenue neutrality would depend on the accuracy of sales forecasts. If it proves inexpensive to increase fuel economy and manufacturers greatly exceed the targets, there would be a net payment from the government to the automotive industry. Conversely, if it is very costly to increase fuel economy and manufacturers fall well short of the targets, there would be a net payment from the industry to the government.

Like tradable credits, feebates would provide direct incentives for all manufacturers to increase the fuel economy of their vehicles.[22] However, unlike the tradable credits system, particular fees and rebates would always be determined by legislation and would not be influenced by market conditions.

Attribute-Based Targets

Figure 5-1 illustrates the principles behind the current CAFE standards. Each dot represents a specific passenger car model—for example, the four-cylinder Accord and the six-cylinder Accord are separate dots. Only those car models that sold at least 1,000 vehicles per year in the United States are shown. The vertical axis shows fuel consumption measured not in miles per gallon but in the amount of gasoline each car needs to travel 100 miles—for example, 25 mpg implies 4.0 gallons to drive 100 miles. The dark horizontal line shows the current CAFE standards: It is placed at 27.5 mpg, which is 3.64 gallons per 100 miles on the vertical axis.

The horizontal axis shows the weight of the car. Cars on the right-hand end weigh more and consume more fuel: They are above the dark CAFE line, which means they are consuming more fuel than the average allowed by the standards. To get back into compliance, a manufacturer that sells heavy cars must also sell some light cars—the left-hand end of the graph. Those cars consume less fuel than the standard. If the manufacturer sells enough light cars, it can produce a fleet average that complies with the CAFE standard of 27.5 mpg.

When the CAFE standards were first implemented, the average car was considerably heavier than today's cars. To comply with the new standard, manufacturers had a considerable incentive to downweight their very largest cars, to produce more small cars, and to encourage consumers to buy those small cars. Thus one effect of CAFE was to reduce the average weight of cars, and this had an undesirable side effect on safety.

There were also equity effects across manufacturers. Some manufacturers had a product mix that emphasized small and medium-weight cars—these manufacturers found it cheap and easy to meet the CAFE standards. Other manufacturers were producing a mix that was more toward the right-hand end of the curve—those manufacturers had to spend a considerable amount of money to develop and sell lighter cars so they could create enough CAFE credits to bring them into compliance with the standards.

These problems arise because the CAFE standards hold all cars to the same fuel economy target regardless of their weight, size, or load-carrying capacity. This suggests that

---

[19]Regulations could be established that require the prices of credits to be made public. Absent such regulations, manufacturers might include confidentiality provisions in their agreements to buy or sell credits.

[20]The tradable fuel economy credits system would differ from tradable credits for sulfur emissions in that the proposed system includes a safety valve, whereas the sulfur emissions system does not. One reason for the difference is that exercise of market power for sulfur emissions credits is much less likely than would be the case for tradable fuel economy credits.

[21]For a more comprehensive study on feebates, see Davis et al. (1995).

[22]There is a duality between feebates and tradable credits; if the targets are identical, the feebate system is linear with respect to deviations in fuel consumption rates from the targets, and the average fuel consumption rates are above the average target. In this case, feebates and tradable credits have identical incentives. They differ from each other if the average fuel consumption rates are below the average targets.

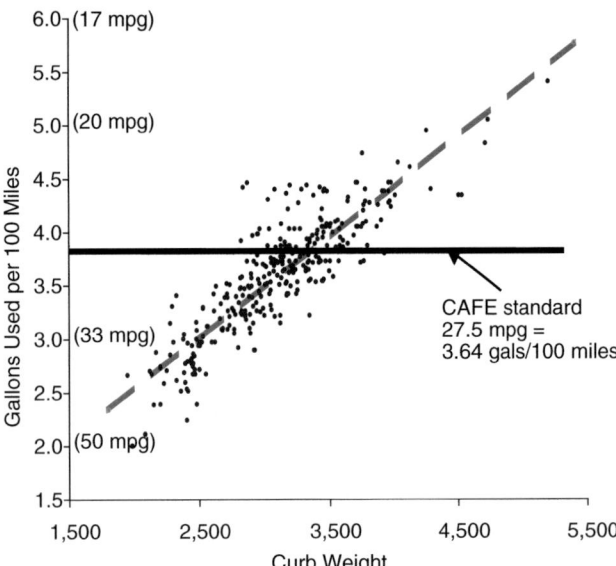

FIGURE 5-1 The operation of the current CAFE standards: passengers cars, gasoline engines only, 1999.

consideration should be given to developing a new system of fuel economy targets that responds to differences in vehicle attributes. For example, the standards might be based on some vehicle attribute such as weight, size, or load. If such attribute-based targets were adopted, a manufacturer would still be allowed to average across all its new vehicle sales. But each manufacturer would have a different target, one that depended upon the average size of the criterion attribute, given the mix of vehicles it sold.[23]

A tradable fuel economy credits system, as described above, could be implemented in combination with the attribute-based targets. The choice of method for setting vehicle economy targets could be separate from choice of incentives to meet targets.

In the current fleet, size, weight, and load-carrying capacity are highly correlated: large cars tend to be heavier, to have room for more people, and to have more trunk capacity than small cars. Choice of a particular attribute as the basis for CAFE measurement will result in incentives for engineers to design vehicles with new combinations of the attributes and to respond to incentives by further varying that particular attribute.

An attribute-based system might use vehicle weight as the criterion. The dashed, upward sloping line in Figure 5-1 shows the average relationship between vehicle weight and fuel consumption. A weight-based CAFE system would use that upward sloping line as its target rather than the current horizontal line.[24]

While a weight-based CAFE system has a number of attractive features, it also has one major disadvantage: It removes incentives to reduce vehicle weight. Judging by recent weight and profit trends, it seems likely the result would be an increase in the proportion of very large vehicles, which could cause safety problems as the variance in weight among vehicles increased. It could also cause an increase in fleet-wide fuel consumption. (These issues are discussed at more length in Attachment 5A.)

Figure 5-2 illustrates an alternative that combines most of the desirable features of the current CAFE standards and the weight-based standard. The target for vehicles lighter than a particular weight (here, 3,500 lb) would be proportional to their weight (e.g., the dashed line in Figure 5-1). But to safeguard against weight increases in heavier vehicles, the target line turns horizontal. Cars heavier than this weight would be required to meet a target that is independent of their weight. (The details of positioning the lines are discussed in Attachment 5A).

These targets provide a strong incentive for manufacturers to decrease the weight of heavier cars—and even a small incentive to increase the weight of the lightest cars. The safety data suggest that the combined effect would be to enhance traffic safety. Accordingly, the committee has named it the Enhanced-CAFE standard (E-CAFE). The Enhanced-CAFE standard may be calibrated separately for cars and for trucks, or it is possible to create a single standard that applies to both types of vehicles, thereby removing the kinds of manipulation possible under the current dual classification system.

The committee views the Enhanced-CAFE system as a serious alternative to the current CAFE system. It holds real promise for alleviating many of the problems with the current regulations. Attachment 5A presents a full description and analysis.

Uniform Percentage Increases

Another possible change would be to require each manufacturer to improve its own CAFE average by some target, say 10 percent; this is often referred to as the uniform percentage increase (UPI) standard. Thus, a manufacturer that was now right at the 27.5 mpg CAFE standard would have to improve its performance to 30.25 mpg. A manufacturer that

---

[23]The manufacturer could average actual gallons per mile and compare that average with the average of target gallons per mile. Alternatively, the manufacturer could average deviations, plus and minus, between actual gallons per mile and target gallons per mile. Whether averaging is done first and deviation calculated second or deviations are calculated first and averaging is done second is mathematically irrelevant.

[24]These possible weight-based targets do not begin to exhaust the possibilities. Many alternative weight-based targets could be designed, or the targets could be based on load-carrying capacity, interior volume, exterior volume, other utility-related attributes, or a combination of these variables (e.g., weight and cargo capacity). The committee did not try to identify and analyze all such possibilities—that would have been well beyond the scope of this study.

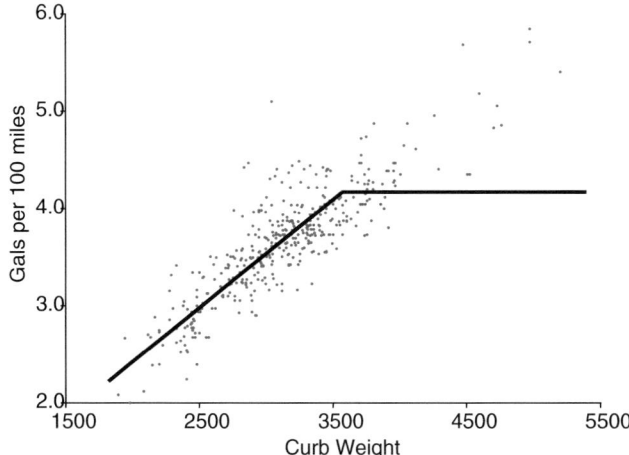

FIGURE 5-2 Fuel economy targets under the Enhanced-CAFE system: cars with gasoline engines, 1999.

was exceeding the current standard at, say, 33 mpg would have to improve its performance to 36.3 mpg.

The UPI system would impose higher burdens on those manufacturers who had already done the most to help reduce energy consumption. The peer-reviewed literature on environmental economics has consistently opposed this form of regulation: It is generally the most costly way to meet an environmental standard; it locks manufacturers into their relative positions, thus inhibiting competition; it rewards those who have been slow to comply with regulations; it punishes those who have done the most to help the environment; and it seems to convey a moral lesson that it is better to lag than to lead.

In addition to fairness issues, the change would not eliminate the problems of the current CAFE system but would create new ones. Implementation of such rules provides strong incentives for manufacturers to not exceed regulatory standards for fear that improvements will lead to tighter regulations. Thus, such rules tend to create beliefs counterproductive for longer-term goals.

Adopt Energy Demand-Reduction Policies

Several alternatives would be aimed more broadly at reducing total gasoline consumption or at reducing all fossil fuel consumption, not simply at reducing the per-mile gasoline consumption of new vehicles. Alternatives include gasoline taxes, carbon taxes, and a carbon cap-and-trade system. Either gasoline taxes or carbon trading/taxes might be part of a comprehensive national energy policy. If these more broadly based policies were implemented, policies aimed directly at fuel economy of new cars might be used along with the broadly based policies, or they could be used in place of one another.

The committee did not devote much time to discussing carbon trading, carbon taxes, and fuel taxes. This does not imply that it considers these options to be ineffective or inappropriate. In fact, such policies could have a much larger short-term and mid-term effect on fuel consumption and greenhouse gas emissions than any of the other policies discussed in this report. The committee did not address these policies comprehensively because they were not part of its charge; instead, it presents here a basic, though incomplete, discussion of these options.

Gasoline Taxes

One alternative, addressed directly at gasoline use, would be an increase in the federal excise tax on gasoline from its current level of $0.184/gal.[25] Every $0.10/gal increase in the gasoline tax would increase the price of gasoline by almost as much.[26]

Increasing the gasoline tax would encourage consumers to drive more efficient vehicles. This would indirectly provide incentives to the manufacturers to increase the fuel efficiency of their vehicles. In addition, a gasoline tax would have an immediate broad impact on gasoline consumption: It would encourage consumers not only to buy more efficient new vehicles but also to drive all vehicles less. If the policy goal is to reduce gasoline consumption and the environmental and oil market impacts of gasoline consumption, then a gasoline tax increase would broadly respond to that goal.

Gasoline taxes, however, have faced significant opposition. Critics point out that gasoline taxes fall particularly hard on rural families and those in more remote locations, where long-distance driving is a normal part of life. It is often asserted that gasoline taxes are regressive and impact the lowest income families the most, even though urban poor and wealthy people typically spend a smaller portion of their income on gasoline than do middle-class families. If the federal gasoline tax were increased, Congress could make the tax revenue neutral or could take other measures to ensure that the change would not cause undue harm.

Carbon Taxes/Carbon Cap-and-Trade Systems

To address problems of global greenhouse gas release, the United States could (1) impose a carbon tax or (2) adopt a carbon cap-and-trade system, under which the total annual emissions of carbon dioxide would be capped or limited to some policy-determined level.

In a system of carbon taxes, each fossil fuel would be

---

[25]In addition, state excise taxes average $0.20/gal, according to the Energy Information Administration (2000).

[26]The price of gasoline would increase by slightly less than the increase in the gasoline tax because the imposition of the tax would reduce oil demand, which in turn would reduce crude oil price and would reduce the per-gallon earnings of refiners and marketers. However, the price and earnings reductions would be small.

taxed, with the tax in proportion to the amount of carbon (or CO_2) released in its combustion. Thus, coal would have the highest tax per unit of energy, petroleum and petroleum products would have a lower tax, and natural gas would have the lowest tax.

In a carbon cap-and-trade system, an energy-producing or -importing firm would be required to possess carbon credits equal to the amount of carbon emitted from its products.[27] A limit would be set on the overall U.S. emissions of CO_2 or of greenhouse gases. Credits could be auctioned off to firms or could be given to firms in proportion to their historical sales of products, and these rights would be tradable among firms. Either way, the price of energy would increase, with the increases on a fuel-by-fuel basis roughly in proportion to the amount of CO_2 released from combustion of the fuels.[28]

A carbon tax, or carbon cap-and-trade system, would have all the advantages of a gasoline tax, and it would extend them to other sectors as well: The rise in power costs would encourage consumers to buy more efficient furnaces, air conditioners, and appliances and would encourage them to use their existing furnaces, air conditioners, and appliances more efficiently. Carbon taxes or trading would provide a broad-based incentive to use less of all fossil fuels, especially those that are particularly carbon-intensive, extending the principle of "least cost" to the entire economy. Implementing such a plan would prevent one energy-using sector from having marginal costs of carbon reduction grossly out of line with those of the other sectors and thus would be an economically efficient means of promoting the reduction of carbon emissions.

Pursue Cooperative Government/Industry Technology Strategies

The final class of strategies would attempt to advance technologies that create dramatic changes in available automotive technologies, changes that could, for the indefinite future, greatly alter fuel economy or types of fuels used. The committee has not evaluated these technology incentive strategies (such an evaluation is beyond the scope of its charge), but it believes such strategies would best be viewed as complementary to the other policy directions, at least once major technological successes had been achieved: A successful technology strategy could greatly reduce the costs of increasing fuel economy. However, in the near term, aggressive requirements to increase fuel economy could divert R&D expenditures away from such technology development efforts and could lead to short-term modest increases in fuel economy at the expense of long-term, dramatic reductions in fuel use.

Partnership for a New Generation of Vehicles

A particular ongoing example of such strategies is the Partnership for a New Generation of Vehicles (PNGV) program. PNGV is a private-public research partnership that conducts precompetitive research directed at new automotive technology. One of its goals is to create marketable passenger cars with fuel economies up to 80 mpg. Each participating company has developed a concept car that would approach the fuel economy goal but that, because of the remaining technological challenges and high production costs, is unlikely to be marketable in the near term. This program has led to advances in technologies such as compression ignition engines, hybrid vehicles, batteries, fuel cells, light engine and vehicle body materials, and advanced drive trains.

Technology Incentives

During the committee's open information-gathering meetings, representatives of the automobile manufacturers proposed that the government provide tax credits to consumers who purchase vehicles that embody new, high-efficiency technology, especially hybrid electric vehicles. Such rebates would strengthen incentives for manufacturers to pursue advanced technology research and then bring those new technologies to market and for consumers to buy these products. However, the committee has not evaluated the particular policy instruments; that evaluation is beyond the scope of this study.

ANALYSIS OF ALTERNATIVES

Dimensions for Assessing Alternatives

In analyzing the alternatives to CAFE, it is important to assess several issues associated with each particular policy instrument as well as the trade-offs among these issues:

- Fuel use responses encouraged by the policy,
- Effectiveness in reducing fuel use,
- Minimizing costs of fuel use reduction,
- Other potential consequences
  —Distributional impacts
  —Safety
  —Consumer satisfaction
  —Mobility
  —Environment
  —Potential inequities, and
- Administrative feasibility.

These are discussed in varying depth below.

---

[27] See, for example, Kopp et al. (2000).

[28] Since all prices would further adjust in response to the changing demand conditions, the final equilibrium price increases would not be exactly in proportion to the amount of CO_2 released but would depend on the various supply and demand elasticities.

Alternative Policies and Incentives for Fuel Use Responses

The various policy measures encourage at least seven quite different fuel use responses. The measures differ in terms of which possible responses they will motivate: those related to the fuel economy of new vehicles, to the usage of vehicles, or to fleet turnover.

Three fuel use responses are directly related to the fuel economy of new vehicles:

- The number of cars sold in each weight class determines the average weight of the new vehicle fleet. All else equal, lighter vehicles use less fuel. This adjustment is referred to as "weight."
- Engine power, other performance characteristics, and vehicle utility can vary. All else equal, greater performance implies more fuel consumed. This adjustment is referred to as "performance."
- More energy-efficient technologies can be incorporated into the engine, drive train, tires, and body structure, often at increased costs, to reduce fuel use without changing weight or performance. This adjustment is referred to as "technology."

Three fuel use responses are directly related to the usage of vehicles:

- The total vehicle miles traveled (VMT) by the entire fleet of vehicles—new and old—is fundamental to fuel consumption. All else equal, fuel consumption is directly proportional to total VMT. This adjustment is referred to as "VMT."
- Driving patterns can be altered—driving during congested times, relative usage frequency of various vehicles, speed. Such adjustment possibilities are referred to as "existing vehicle use."
- Maintenance of existing vehicles can influence fuel use. All else equal, a well-maintained vehicle will use less fuel. This adjustment is referred to as "vehicle maintenance."

One possible fuel use response is directly related to vehicle fleet turnover:

- The rate of retirement of older, generally less fuel-efficient vehicles influences the rate at which the overall fuel economy of the vehicle fleet changes, especially if new vehicles are more fuel-efficient than old vehicles. This adjustment is referred to as "vehicle retirements."

Effectiveness in Reducing Fuel Use

The various policy instruments differ in which responses they motivate and therefore differ in terms of their effectiveness in reducing fuel use. Table 5-1 summarizes fuel use responses caused by various policy instruments. It provides a very rough indication of whether a particular policy instrument would lead to a larger adjustment than would other instruments. Each cell of the table indicates changes relative to what would be the case absent the particular policy being considered, with all else held equal. It should be noted that for these comparisons, the magnitude of responses associated with any particular policy instrument depends on how aggressively that instrument is implemented—the severity

TABLE 5-1 Incentives of the Various Policy Instruments for Seven Types of Fuel Use Response

Fuel Use Adjustment Policy Instrument	Average Weight	Performance	Technology	VMT	Existing Vehicle Use	Vehicle Maintenance	Vehicle Retirements
Current CAFE, with or without two-fleet rule	Decrease weight	Reduce performance	Incorporate more fuel-efficient technology	Small increase in VMT	No impact on existing vehicle use	No impact on vehicle maintenance	Insignificant impact on vehicle retirements
CAFE with reformed car/truck differentiation							
Current CAFE targets with tradable credits							
Current CAFE but with E-CAFE weight-based targets	Reduce weight of heavy vehicles; increase weight of light vehicles						
Tradable credits E-CAFE weight-based targets							
UPI for each manufacturer							
Feebates							
Carbon taxes/ carbon trading	Decrease weight			Decrease in VMT	Shift to reduce fuel use	Improve maintenance	Increase retirements of low fuel economy vehicles
Gasoline taxes							

and speed of the intervention—and on the choices the manufacturers and the consumers make in responding to the implementation. Table 5-1 incorporates an assumption that severity and implementation speed of the interventions are similar for each instrument.

Table 5-1 indicates that neither CAFE nor the alternative instruments directed at fuel economy (feebates, tradable credits, weight-based targets, tradable credits with weight-based targets, UPI for each manufacturer) have any significant gasoline-reducing impacts through VMT, existing vehicle use, vehicle maintenance, or old vehicle retirements. Only the two broad-based demand reduction policies—carbon taxes/trading or gasoline taxes—provide incentives for people to drive less, to use their more fuel-efficient vehicle when possible, to maintain existing vehicles, or to retire old fuel-inefficient vehicles.

Any of the instruments directed toward new car fuel economy would result in small increases in VMT, usually referred to as the rebound effect, because increasing the fuel economy of new vehicles decreases the variable cost per mile of driving, encouraging consumers to drive more. As discussed in Chapter 2, empirical studies suggest that every 10 percent increase in gasoline price leads to a 1 or 2 percent reduction in vehicle miles. Similarly, a 10 percent increase in fleet economy can be expected to lead to a 1 or 2 percent increase in VMT. The net impact would be an 8 or 9 percent reduction in fuel use for every 10 percent increase in fuel economy of the entire vehicle fleet.

Table 5-1 indicates that all the policy instruments will motivate changes in technology that reduce gasoline use, and they all motivate reductions in performance. Each instrument, other than those including proportionate weight-based targets, may reduce fuel use through reductions in the average vehicle weight. Proportionate weight-based targets, on the other hand, are unlikely to motivate weight reductions, since reducing vehicle weights would not bring the manufacturer closer to the targets.

The broad demand-reduction policies differ from the other policies in the way they motivate changes in weight, power, and technology. CAFE and the alternative instruments directed at fuel economy are felt directly by the manufacturer and give it a direct incentive to offer lighter, lower-performance cars with more energy-efficient technologies. Gasoline taxes and carbon taxes motivate consumers to shift their new vehicle purchases, directing them toward lighter, lower-performance cars with more energy-efficient technologies, which, in turn, provides a stimulus to the manufacturers.

In principle, a direct incentive for manufacturers to sell vehicles with increased fuel economy and an equally strong incentive for consumers to buy vehicles with increased fuel economy should have similar fuel economy impacts. Manufacturers would change the characteristics of their vehicles in anticipation of changes in consumer choices caused by higher fuel taxes. Manufacturers facing a direct incentive would respond to that incentive and would pass the financial incentives on to the consumers, who would modify their choice of vehicles.

In practice, however, considerable uncertainty is involved in calibrating a gasoline tax or carbon tax so that it would have the same impact as a direct incentive for manufacturers. If consumers discounted future gasoline costs over the entire vehicle life, using identical discount rates, a gasoline tax and a direct manufacturer incentive would change fuel economy to the same extent if the manufacturer incentive were equal to the discounted present value of the gasoline-tax incentive. However, if some new vehicle buyers count, say, only 3 years of gasoline costs, the discounted present value of the gasoline tax cost over 3 years would provide an equivalent manufacturer incentive. The results of the two calculations differ by a factor between two and four, depending on the discount rate used,[29] so there is significant uncertainty about how large a gasoline tax would, in fact, be equivalent to a direct manufacturer incentive.

Minimizing the Costs of Fuel Economy Increases

A fuel economy policy may produce an increase in manufacturing cost through changes that require more costly materials, more complex systems, more control systems, or entirely different power train configurations. A fuel economy policy may also impose nonmonetary costs on consumers through reductions in vehicle performance or interior space. Performance reductions may make vehicles less attractive for towing trailers, for merging quickly onto freeways, or for driving in mountainous locations.

Next, the committee examines whether particular policy instruments would minimize the overall cost of whatever fuel reductions are achieved. This is a matter both of the structure of the policy instruments and of the aggressiveness of policy implementation, including the lead time manufacturers have to make the requisite adjustments. Also examined is whether the policy instruments can be expected to reach an appropriate balance between higher vehicle costs and the savings from reduced gasoline use (including external costs of fuel use).

Each issue of cost-minimizing policy and cost-minimizing manufacturer and consumer response can be examined based on most likely costs. However, no one has perfect

---

[29]Consider a gasoline tax intended to embody the same incentive as a tradable credit with a credit price of $0.26/gal over a 150,000-mile lifetime. Under tradable credits, a vehicle with a fuel economy of 25 mpg when the standard was 30 mpg would need $150{,}000 \times [(1/25) - (1/30)]$ credits, or 1,000 credits, with a value of $260. Assume the buyer discounts gasoline use over only the first 3 years, at a 10 percent discount rate, driving 15,000 miles per year. A gasoline tax of $1.04/gal would give a discounted tax difference of $260 between a 30-mpg and a 25-mpg car, the same as the value of the credits. Thus, for such a consumer, a manufacturer incentive of $0.26/gal for each of 150,000 miles is economically equivalent to a $1.04/gal gasoline tax.

knowledge about all costs and no one can predict with confidence changes in technology or economic conditions. Therefore, the discussion focuses on whether various policy instruments allow manufacturers, consumers, and policy makers to respond to uncertainty in future conditions.

This section provides a conceptual framework for examining these questions.

Minimizing the Total Cost of Fixed Economy Increases

The issue of minimizing the total cost to society of whatever average fuel economy increases are achieved depends critically on the observation that vehicles differ significantly from one another in size, weight, body type, and features, and that some manufacturers offer a full line of such vehicles and others offer only a limited one. To minimize the overall cost to society of a given reduction in fuel use, the marginal cost of reducing gasoline use must be the same (or approximately the same) across all vehicles.[30] Manufacturers must be able to meet this condition for all vehicles, including domestic and import fleets of one manufacturer. Thus, under ideal conditions, policy options should not induce significant differences in the marginal costs among manufacturers. Policies that do induce significant differences will not minimize the total cost to society of whatever level of average fuel economy increase is achieved.

Balancing the Costs and Benefits of Reducing Fuel Use

Two components, private costs (and benefits) and external costs, must be considered when evaluating costs and benefits of changing fuel use. The private costs of the vehicle and its operation, including the fuel costs, can be estimated by the new vehicle purchaser through reading magazines, government reports, or examining the stickers on the car or truck being considered for purchase. It is in the purchaser's own interest to take into account these private costs, not just for the time he or she expects to own the vehicle but also for periods after it is sold as a used vehicle, since the resale value can be expected to reflect these gasoline purchase costs. The manufacturer can adequately balance this portion of costs against the manufacturing costs of reducing fuel use.

Accounting for external costs is more difficult. Absent policy intervention, it is not in the owner's own interests to account for these costs. These costs must be incorporated into financial incentives facing consumers or manufacturers to balance the costs and benefits of reducing fuel use: The efficient policy intervention would require the manufacturer or consumer to face a financial incentive that added a cost equal to the external marginal cost of additional gasoline used. If the incentive were substantially greater than the marginal external cost, then the intervention would be too severe—the fuel savings would cost more than they were worth. If the incentive were substantially smaller than the marginal external cost, then the intervention would be too weak—the additional fuel savings would cost substantially less than they were worth. Thus, given the magnitude of external costs, one can determine the appropriate severity of incentives.

This task is complicated by the uncertainty facing the policy process. Although future gasoline prices can be estimated, they cannot be predicted with certainty (EIA, 2000). Estimates of external costs of gasoline use have wide error bands. In addition, estimates of the marginal cost of fuel use reductions will remain quite uncertain until these costs are engineered into vehicles. Therefore, in evaluating any particular policy instrument, it is important to assess how the instrument would operate given real-world uncertainties.

The task is also complicated by the existing gasoline taxes. The consumer currently faces $0.38/gal combined state and federal excise taxes on gasoline as part of the gasoline price. These taxes are costs from the perspective of the consumer but are transfer payments from the perspective of the nation. For consumer decisions about how much to drive, these gasoline taxes can be seen as user fees covering the costs of building and maintaining roads, highways, and associated infrastructure and providing services to motorists. However, for consumer decisions about fuel economy of new vehicles, the user service interpretation is inappropriate. As discussed above, the committee has identified a total external marginal cost associated with the fuel economy of new vehicles of about $0.26/gal, based on CO_2 emissions and international oil market impacts. But the price the consumer faces in purchasing gasoline already includes an average of $0.38/gal, an amount larger than the committee's estimate of those external costs that vary with fuel economy of new vehicles. Thus, if existing gasoline taxes are included, consumers are already paying tax costs larger than the external costs seen as justifying intervention to increase fuel economy above the market-determined levels. If existing gasoline taxes are excluded, perhaps because they are used primarily as a mechanism to collect revenue for states and for the federal government, then the $0.26/gal remains as a reason for intervening so as to increase fuel economy beyond market-determined levels.

---

[30]This equal-marginal-cost rule is easiest to see by considering an outcome for which different vehicles had very different marginal costs of reducing fuel use. For example, assume that the cost for vehicle A of reducing lifetime fuel use by 200 gallons is slightly more than $300 for the next 200 gallons reduction or slightly less than $300 for the previous 200 gallons (that is, about $1.50/gal) and assume that the cost for vehicle B of reducing lifetime fuel use by 200 gallons is slightly more than $500 for the next 200 gallons or slightly less than $500 for the previous 200 gallons (that is, about $2.50/per gal). In that situation, reducing vehicle A lifetime gasoline use by 200 gallons and increasing vehicle B lifetime gasoline use by 200 gallons would not change the total lifetime fuel use for the two cars but would reduce total costs by about $200, the difference between the saving for vehicle B and the additional cost for vehicle A.

TABLE 5-2 Issues of Cost Minimization for the Various Policy Instruments

Policy Instrument	For Manufacturer, Is Marginal Cost of Fuel Reduction Same Across Fleet of Cars and Trucks?	For Manufacturer, Is Marginal Cost of Fuel Reduction Same Across Domestic Fleet and Import Fleet?	Is Marginal Cost of Fuel Reduction Same Across All Manufacturers?	Can Policy Assure Balance of Costs and Benefits with Uncertain or Changing Economic Conditions?
Current CAFE, with two-fleet rule	No. Probably significant differences	Unlikely	No. Probably significant differences	No
Current CAFE, without two-fleet rule		Yes		
CAFE with reformed car/truck differentiation		Only if two-fleet rule removed		
Current CAFE targets with tradable credits	Yes	Yes	Yes	Yes
Current CAFE but with E-CAFE weight-based targets	No. Probably significant differences	Only if two-fleet rule removed	Better than current CAFE	No
Tradable credits, E-CAFE weight-based targets	No	Yes	Yes	Yes
UPI for each manufacturer	Same as CAFE	Same as CAFE	No	No
Feebates	Yes	Yes	Yes	Yes

Analysis of the Costs of Various Policy Instruments

With that framework in place, the policy instruments can be analyzed in terms of cost minimization for a given fuel economy increase and balancing of marginal costs of fuel economy increases with marginal benefits. Table 5-2 provides a summary.

Several of the policy instruments satisfy all rules for minimizing the cost for obtaining a given mean increase in fuel economy: feebates, gasoline taxes, carbon taxes, and tradable credits with attribute-independent or with proportionate[31] weight-based targets. Therefore, each of these instruments would minimize the cost of obtaining a given mean fuel economy increase. Furthermore, gasoline taxes and carbon taxes would roughly minimize the overall cost of reducing fuel use, not simply of increasing fuel economy, since they also include incentives to reduce VMT. In addition, each of these instruments (feebates, tradable credits, gasoline taxes, or carbon taxes) could efficiently balance the costs and benefits of fuel economy even under uncertain or changing economic conditions, if the incentives in these are chosen correctly. This would require the marginal incentives to equal the marginal value of the external costs (in our example, $0.26/gal of gasoline). At the other end of the spectrum, two of instruments—the current CAFE targets and UPI for each manufacturer—satisfy none of the conditions unless the two-fleet rule is eliminated and the distinction between passenger cars and trucks is appropriately reformed. Elimination of the two-fleet rule and elimination or appropriate repair of the distinction between passenger cars and trucks would improve both systems. Even if the two-fleet rule is eliminated and the car/truck distinction appropriately repaired, neither of these instruments can be expected to lead to an efficient allocation of reductions among manufacturers or to an economically efficient overall level of fuel economy increases.

Performance Trade-offs for the Various Policy Instruments

The discussion above makes it clear that there are trade-offs among the various performance objectives important for the various policies. In particular, there may be trade-offs between three separate performance objectives:

- Flexibility of choice for manufacturers and consumers to allow appropriate fuel economy increases, while maintaining consumer choice;
- Certainty, or predictability, in the magnitude of total fuel consumption reductions, or at least certainty in the magnitude of new vehicle fuel economy increases; and
- Effectiveness in motivating the broadest range of appropriate consumer and manufacturer fuel use responses, and possibly the greatest fuel use reductions.

Table 5-3 summarizes these trade-offs. For each policy instrument it provides a summary of performance of the instrument in terms of these three objectives and compares them where possible. Each summary of the performance characteristics is stated relative to the current CAFE system.

Since none of the policy instruments provides certainty about the magnitude of fuel use reductions, primarily because

---

[31]The addition of the enhanced-CAFE weight-based targets, even with tradable credits, would give different incentives for increasing fuel economy for heavy and light cars.

TABLE 5-3 Performance Trade-offs for the Various Policy Instruments

Policy Instrument	Will Manufacturers and Consumers Have Flexibility of Choice to Allow Economically Efficient Fuel Economy Increases?	How Certain Will Be the Magnitude of New Car Fuel Economy Increases?	How Broad a Range of Fuel Use Responses Would Be Motivated?
Current CAFE, with two-fleet rule	No	Much certainty	Broad range of fuel economy responses
Current CAFE, without two-fleet rule	No, but more flexibility than current CAFE		
CAFE with reformed car/truck differentiation	Similar to current CAFE	More certainty than current CAFE	
Current CAFE targets with tradable credits	Yes	Less certainty than current CAFE	Broader than current CAFE
Current CAFE but with E-CAFE weight-based targets	More flexibility than current CAFE		
Tradable credits, E-CAFE weight-based targets	Yes		
UPI for each manufacturer	Similar to current CAFE	Similar to current CAFE	Broader range than current CAFE
Feebates	Yes	Less certainty than current CAFE	Broader range than current CAFE
Gasoline taxes		Considerably less certainty than current CAFE	Broader than current CAFE, plus VMT responses
Carbon taxes			Like gasoline taxes, plus economy-wide responses

none can directly control total vehicle miles, the second column of Table 5-3, "How certain will be the magnitude of new car fuel economy increases?", summarizes the degree of certainty in the magnitude of new vehicle fuel economy increases not in the magnitude of total fuel use reductions.

The third column, "How broad a range of fuel use responses would be motivated?", summarizes the material from Table 5-1. Broader ranges of responses tend to lead to larger responses for a given magnitude of incentives and thus larger expected reductions in fuel use.

Table 5-3 underscores the observation that choice among instruments will necessarily involve trade-offs among those three performance objectives. In particular, there is an inherent trade-off between flexibility and certainty. Generally, policy instruments that give the most certainty of response are the ones that allow the least flexibility of choice. In addition, certainty of response should not be confused with magnitude of response. For equivalent magnitudes, the broad energy demand reduction policies—gasoline taxes and carbon taxes—have less certainty of magnitude but can be expected to lead to the greatest reductions in fuel use.

In addition to the trade-offs highlighted in Table 5-3, the various policy options will have additional impacts and therefore additional trade-offs. The following section of this chapter points out some of the additional impacts of the various policy options.

Table 5-3 incorporates an implicit assumption that the severity and implementation speed of the interventions are similar for each instrument. However, this analytic requirement for consistency is not a requirement for policy making. For example, it is quite possible to decide whether one would prefer to implement a system of tradable credits with aggressive fuel economy targets or to increase the targets of the CAFE system less aggressively. Or, one could choose between a moderate carbon tax system or an aggressive gasoline tax increase. Further analysis of the trade-offs involved in such choices is a necessary part of the policy-making process.

Other Potential Consequences

In addition to issues discussed above, there may be other consequences, some unintended, associated with these instruments. Six classes of potential consequence have been identified: distributional impacts, automotive safety, consumer satisfaction, mobility of the population, employment, and environmental impacts. These will be discussed in what follows.

Distributional Impacts of the Instruments

In addition to the economic efficiency impacts of the various policy instruments, there are also impacts on the distribution of income/wealth between automotive companies, consumers, and the government. Carbon taxes and gasoline taxes could have the greatest distributional consequences: Vehicle owners would pay taxes to the federal government even if the fuel economy of their cars or trucks exceeded the fuel economy targets. For example, with a $0.26/gal gasoline tax, the consumer driving a 30-mpg vehicle over a 150,000-mile lifetime would pay $1,300 in gasoline taxes during the life of the vehicle. The payment to the government would allow other taxes to be reduced or would allow additional beneficial government spending.

Current CAFE standards (or UPI), with or without the two-fleet rule, with or without a redesign of the distinction between cars and trucks, and with or without attribute-based targets, would not lead to any financial transfer among manufacturers who meet or exceed the targets. However, if, to meet the targets, those manufacturers need to increase the price of low-mpg vehicles and decrease the price of high-mpg vehicles, then this pricing strategy would have a differential impact on consumers. In addition, those manufacturers who fail to meet the targets would pay financial penalties to the government.

Like the current CAFE systems, feebates and tradable fuel economy credits cause no financial transfers for those manufacturers just meeting the target. Like the current CAFE systems, both systems would lead to payments from those manufacturers who failed, on average, to meet the targets. A difference is that, unlike the current CAFE system, those manufacturers exceeding the fuel economy targets receive a financial transfer and that transfer provides the motivation for further increases in fuel economy.

Feebates would require manufacturers with fuel economy lower than the targets to pay money to the federal government, while those with higher fuel economy would receive payments from the government. Although money would pass through the government, on net there would be a transfer of payments between the two groups of manufacturers. If the mean fuel economy of the entire new vehicle fleet exceeds the mean target, there would be a financial transfer from the government to the automotive industry as a whole; conversely, there would be a financial transfer from the industry to the government if the mean economy is lower than the average target. Thus, whether the system adds net tax revenues or subtracts from tax revenues would depend on the average fuel economy of the new cars sold.

Tradable credits, either with or without weight-based targets, could lead to financial transfers between automobile manufacturers. However, the transfers would be smaller (on a per-vehicle basis) than the transfers associated with an equivalent gasoline tax or a carbon tax. In the example above, with a 30-mpg target and a $0.26/gal price of credits, a manufacturer whose mean economy was 35 mpg would receive $185 per vehicle from selling credits, and the manufacturer of a 25-mpg vehicle would need to pay $260 per vehicle for purchasing credits. These could be compared with the $1,114 per-vehicle consumer payment of gasoline taxes (at $0.26 per gallon) for the 35-mpg vehicle and the $1,560 per-vehicle consumer payment of gasoline taxes for the 25-mpg vehicle.

Safety

As discussed in Chapters 2 and 4, over the last several decades, driving continued to become safer, as measured by the number of fatalities and severe injuries per mile of driving. This trend has resulted from a combination of factors, including improvement in highway design; enforcement of traffic laws, including alcohol restrictions; and improved design of vehicles. Although most of these factors are apt to be unaffected by changes in CAFE standards, one of them, the design of vehicles—in particular, their weights and sizes—may well be influenced by changes in the form and severity of CAFE standards.

The relationships between weight and risk are complex and have not been dependably quantified. However, in general, it appears that policies that result in lighter vehicles are likely to increase fatalities (relative to their historic downward trend), although the quantitative relationship between mass and safety is still subject to uncertainty.

Chapter 4 estimated that about 370 additional fatalities per year could occur for a 10 percent improvement in fuel economy if downweighting follows the pattern of 1975–1984. The downweighting by itself would improve fuel economy by about 2.7 percent for cars and 1.0 percent for trucks, which would reduce gasoline consumption by about 4.3 billion gallons. Thus, if this relationship accurately described future downweighting, and if for policy analysis purposes a benefit of $4 million is assigned to every life, then the safety costs would be $1.4 billion, or $0.33 in costs of lost lives per gallon of gasoline saved by downweighting.[32] If, on the other hand, weight reduction was limited to light trucks, the net result could be a reduction in fatalities, with a safety benefit. Note that there is much uncertainty in the estimate of value of lives and in the number of fatalities. Higher or lower values of these figures would increase or decrease the estimate in safety costs or benefits proportionately.

As discussed above, proportionate weight-based targets would eliminate motivation for weight reductions and thus would avoid any adverse safety implications. Inclusion of the enhanced-CAFE weight-based targets would eliminate any motivation for weight reductions for small vehicles and would concentrate all weight reductions in the larger vehicles, including most light-duty trucks, possibly improving safety.

---

[32]The committee has not determined what fraction of this cost consumers already take into account in their choices of vehicles. If consumers are already taking into account consequences of vehicle purchases for their own safety, these safety consequences help explain consumer preferences for larger cars. If consumers do not take this cost into account, the $0.33 per gallon cost of downweighting overwhelms the $0.26 per gallon reduction in external costs that would result from downweighting. Finally, if consumers already fully include consequences of car purchases for their own safety but ignore all consequences of car purchases for the safety of other drivers with whom they may collide, then these costs imposed on others would represent another external cost, one associated strictly with choice of new vehicle weight. The committee has not quantified this possible source of external costs.

All other policy instruments could be expected to reduce vehicle weights and thus, all else equal, could be expected to have unintended safety impact. Any policy instrument that encourages the sale of small cars beyond the level of normal consumer demand may have adverse safety consequences.

Consumer Satisfaction

Reductions in weight or performance of vehicles below that desired by consumers could reduce vehicle purchaser satisfaction by reducing the utility of the vehicles available to consumers.[33] Reduction in satisfaction might reduce purchases of new vehicles and thus adversely impact the auto industry and employment in that industry. This issue may be particularly important for policies that implicitly or explicitly restrict automotive firms from selling vehicles designed to appeal to consumers whose preferences differ from those of typical customers. For example, some consumers expect to use vehicles for towing boats or trailers, for farming, or for construction, activities in which vehicle power—at relatively low speeds—is more important than fuel economy. Others consumers may desire luxury features that sacrifice fuel economy. Some may use their vehicle for carpooling several families or transporting youth soccer teams, Girl Scout troops, or school groups and thus may need the ability to safely seat many people.

All of the policy instruments provide strong incentives to reduce performance, and many also provide incentives to reduce weight. Such reductions could have unintended negative consequences. Policy instruments that provide no motivation for weight reduction are less likely to have such negative consequences.

Mobility

Personal mobility is highly valued in American society. People living in suburban and rural locations often have no alternatives to light-duty vehicles for personal mobility. Large numbers of people cannot afford to live close to their work, and many families include two wage-earners who may work a significant distance from one another. Thus, any policies that reduce the mobility of these people may create unintended hardships.

Mobility may be reduced by policies that greatly increase the cost of driving. Large enough gasoline taxes or carbon taxes could have this impact. Conversely, any policy that, all else equal, increases fuel economy could reduce the cost of driving and could increase options for mobility.

Employment

Employment in the U.S. economy is linked primarily to monetary and fiscal policies pursued at the federal level and to regional policies that allow a range of employment opportunities throughout the United States. None of the policy options can be expected to significantly impact monetary and fiscal policies. However, policies that reduce the number of vehicles manufactured in the United States or rapidly and significantly reduce the scale of an industry central to a regional economy could have at least temporary regional employment impacts.

The committee believes that none of the policies discussed here would have such negative consequences if implemented wisely. If implemented too aggressively, any of the policies could greatly increase the cost of vehicles or their use and thus have the potential for harming employment. In the remainder of this chapter, it is assumed that no policy will be implemented so aggressively that there would be such employment impacts.

One particular issue was raised in the committee's open sessions. It was suggested that maintenance of the two-fleet rule, requiring each manufacturer's domestic fleet and imported fleet of passenger cars to separately meet the CAFE standards, was important for avoiding job losses in the United States, particularly in the automotive industry. However, the committee found no evidence nor was it offered any evidence or analysis to support that contention. In addition, the current rule does provide a strong motivation for manufacturers to reduce domestic content of some vehicles, particularly the larger vehicles, to keep them in the import-fleet category. Therefore, while at some time the two-fleet rule may have protected domestic production, the committee sees no reason to believe it continues to play this role at all. Therefore, elimination of the two-fleet rule is not expected to have net adverse impacts on employment in the U.S. automotive industry.

Environment

Environmental impacts can be viewed in two categories, depending on whether the impacts are closely related to the amount of gasoline used or are independent of gasoline use but instead are dependent on the VMT and on the characteristics of individual vehicles.

In the first category are the environmental consequences of the release of CO_2, which is directly proportional to the amount of gasoline consumed. These direct environmental externalities, discussed previously, are a major reason for market intervention.

In the second category are the environmental consequences of criteria pollutants emitted from cars and trucks. For a given vehicle, the more it is driven, the greater will be the amount of emissions released and the greater the environmental impact. Thus, policies that reduce VMT—gaso-

---

[33]The effect of weight reduction on consumer satisfaction is complicated, however, because structural redesign and the use of lightweight materials can allow weight reduction without changes in structural integrity or handling.

line taxes and carbon trading/taxes—can lead to additional environmental benefits.

New cars emit far less criteria pollutants per mile driven than do cars of older vintage. Thus, policies that reduce retirements of old cars (by, for example, significantly increasing the costs of new vehicles) would increase the average emissions per vehicle mile traveled and the total emissions of criteria pollutants.

Potential Inequities

The issue of equity or inequity is subjective. However, one concept of equity among manufacturers requires equal treatment of equivalent vehicles made by different manufacturers. The current CAFE standards fail this test. If one manufacturer was positioned in the market selling many large passenger cars and thereby was just meeting the CAFE standard, adding a 22-mpg car (below the 27.5-mpg standard) would result in a financial penalty or would require significant improvements in fuel economy for the remainder of the passenger cars. But, if another manufacturer was selling many small cars and was significantly exceeding the CAFE standard, adding a 22-mpg vehicle would have no negative consequences.

This differential treatment of identical vehicles characterizes the current CAFE system with or without the two-fleet rule, with or without reclassifications of trucks, and without weight-based targets. With the enhanced-CAFE weight-based targets, this differential treatment would continue to exist.

Another notion of equity involves whether manufacturers are rewarded, treated neutrally, or punished for incorporating fuel-economy-enhancing technologies when they are not required by law to do so. Rewards or neutral treatment seem equitable; punishment seems inequitable.

Uniform percentage improvements would operate in this inequitable manner. Consider two initially identical manufacturers initially selling identical fleets of vehicles, both just meeting current CAFE standards. Suppose that one manufacturer of its own volition introduces improved technologies that increase fuel economy and the other does not; suppose further that some years later the government adopts a UPI regulation. The first manufacturer would then be required to achieve a higher fuel economy than the second. But the first manufacturer would already have used the low-cost fuel-saving technologies and would be forced to use higher-cost technologies for the further improvement. That manufacturer would be significantly punished for having improved fuel economy beyond what was required by law.

A final concept of equity among manufacturers is more global. New policies that would impose costs disproportionately on particular manufacturers, who themselves have remained in compliance with existing law and policies, seem inequitable because they would impose unequal costs on otherwise similarly situated manufacturers.

A policy decision to simply increase the standard for light-duty trucks to the same level as for passenger cars would operate in this inequitable manner. Some manufacturers have concentrated their production in light-duty trucks while others have concentrated production in passenger cars. But since trucks tend to be heavier than cars and are more likely to have attributes, such as four-wheel drive, that reduce fuel economy, those manufacturers whose production was concentrated in light-duty trucks would be financially penalized relative to those manufacturers whose production was concentrated in cars. Such a policy decision would impose unequal costs on otherwise similarly situated manufacturers.

Administrative Feasibility

Perhaps the easiest policy to implement would be increases in gasoline or other motor fuel taxes. There is already a system of gasoline taxation in place, and implementation of the policy would simply involve increasing the tax rate. No additional administrative mechanisms would be required.

It is also relatively easy to enforce the current CAFE standards. These standards, with or without the two-fleet rule, involve reporting the sales of the various models of vehicles and reporting the fuel economy of each model. The fuel economy data are already developed for reporting to consumers. The sales data and the associated mathematical calculations could be self-reported. Accounting for carry-forward and penalties is straightforward. Thus, CAFE standards involve little or no administrative difficulty except in setting the standards. However, setting target levels requires significant economic and technological information, so the levels should be revised periodically. Similarly, weight-based targets require only one additional piece of readily available data, the vehicle curb weight. Thus, the administrative issues would be virtually no different from those of the current CAFE standards.

Feebates require financial transfers between the auto manufacturers and the government and thus require a reporting and collecting function. This function could be integrated with other taxation functions but would require administrative efforts.

Tradable credits, with or without weight-based targets, require a new administrative mechanism. Perhaps the easiest approach would involve an extension of the reporting mechanisms required under the current CAFE standards. At the end of the year, the manufacturer would report the number of cars sold and their fuel economies (based on the standard testing procedure) and their weights, if weight-based targets are to be used. They would also report the number of fuel economy credits purchased or sold, the names of the other companies involved in the transactions, and sales prices of credits. The government agency might require public reporting of the credit sales prices to allow use of this information for further rule making. Reporting on the names of the other companies involved in transactions would allow cross-checking to assure that every reported purchase of credits had a corresponding reported sale. The government agency would also need a mechanism to sell credits to any firm that decided to buy them. In general, however, the administrative requirements would be modest.

Introducing a carbon tax or carbon trading mechanism would require a system that extended well beyond just the automobile manufacturers. Discussion of how one might set up such a system is well beyond the scope of this report.

REFERENCES

Davis, W.B., M.D. Levine, and K. Train. 1995. Effects of Feebates on Vehicle Fuel Economy, Carbon Emissions, and Consumer Surplus. DOE/PO-0031. Department of Energy, Office of Policy, February. Washington, D.C.: DOE.

Delucchi, M.A., D.L. Greene, and M. Wang. 1994. "Motor Vehicle Fuel Economy: The Forgotten Hydrocarbon Control Strategy?" *Transportation Research-A* 28A: 223–244.

EIA (Energy Information Administration). 2000. A Primer on Gasoline Prices (an Update). Available online at <http://www.eia.doe.gov/pub/oil_gas/petroleum/analysis_publications/primer_on_gasoline_prices/html/petbro.html>.

EIA. 2001. Alternatives to Traditional Transportation Fuels. Table 13: Estimated Consumption of Alternative Transportation Fuels in the United States, by Vehicle Ownership, 1997, 1999, and 2001 (Thousand Gasoline–Equivalent Gallons). Available online at <http://www.eia.doe.gov/cneaf/alternate/page/datatables/atf1-13_00.html>.

EMF (Energy Modeling Forum). 1982. World Oil–Summary Report. EMF Report 6, Vol. 1, February. Stanford, Calif.: EMF, Stanford University.

Kopp, R., R. Morgenstern, W. Pizer, and M. Toman. 2000. A Proposal for Credible Early Action in U.S. Climate Policy. Washington, D.C.: Resources for the Future. Available online at <http://www.weathervane.rff.org/features/feature060.html>.

Leiby P.N., D.W. Jones, T.R. Curlee, and L. Russell. 1997. Oil Imports: An Assessment of Benefits and Costs. ORNL-6851. Oak Ridge, Tenn.: Oak Ridge National Laboratory.

Stoffer, H. 2001. "CAFE Can Discourage U.S. Employment," *Automotive News*, No. 5921:2.

Attachment 5A

Development of an Enhanced-CAFE Standard

This attachment develops the Enhanced-CAFE (E-CAFE) standard, an alternative to the current fuel economy regulations. This alternative has a number of advantages: It has the potential to decrease fuel consumption, reduce the "gaming" of the fuel economy standards, and increase the safety of the overall vehicle fleet. The committee views the new system as a serious alternative to the current CAFE standards. Because of limitations of time and data, it has only been able to do an approximate calibration of the effects of the new system. Thus, although the E-CAFE standard is highly promising, some additional analysis will be required.

TARGETS UNDER THE CURRENT CAFE STANDARD

Figure 5A-1 shows the general relationship between fuel consumption and vehicle weight for passenger cars, based on 1999 data. Fuel consumption, the vertical axis, is expressed as number of gallons needed to drive 100 miles. Each point in the graph is a single car model, e.g., a four-cylinder Accord. (Car models that sell fewer than 1,000 cars per year in the United States are not shown.)

The current CAFE standard sets a passenger car target of 27.5 mpg (3.64 gallons per 100 miles of driving) for each manufacturer. Compliance is determined by averaging gallons per mile across the manufacturer's entire fleet of cars. With averaging, the manufacturer can produce some cars that get low mpg if it balances them with enough cars that get high mpg. The horizontal line in Figure 5A-1 shows the CAFE target.

Point A in Figure 5A-1 represents a car model that consumes more fuel than is allowed by the CAFE standard. The gap between point A and the horizontal CAFE line is the amount of excess fuel consumption. Point B is a car that consumes less fuel than the CAFE standard. The gap between point B and the CAFE line is not as large, so the manufacturer who makes As and Bs will have to sell approximately two point B cars to offset the high fuel consumption of one point A car. Manufacturers have an incentive to sell more of the lighter

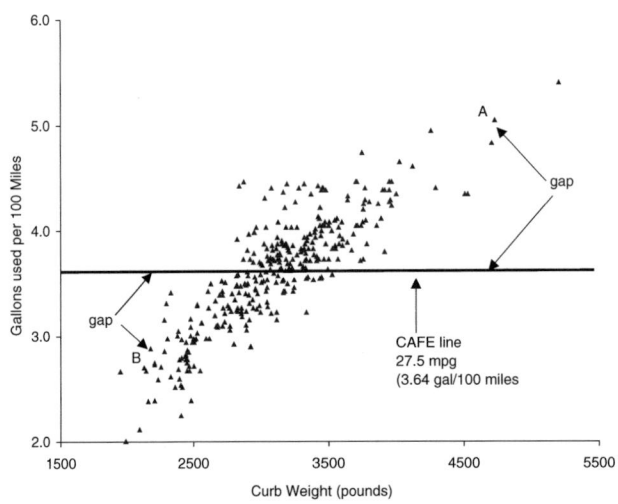

FIGURE 5A-1 Gallons used per 100 miles (cars only, gasoline engines only).

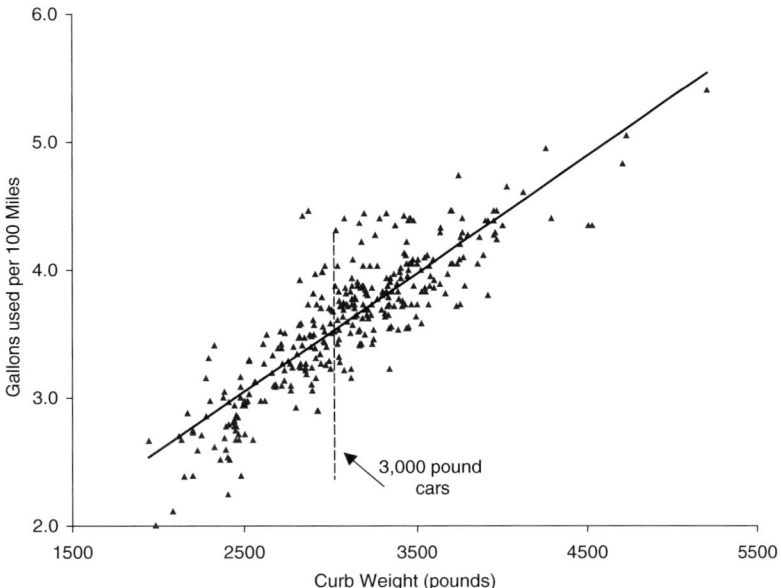

FIGURE 5A-2 Regression line through the car data in Figure 5A-1 (passenger cars only, gasoline engines only).

cars, which may lead to an increase in traffic fatalities. In addition, there are equity problems: Those manufacturers who specialize in making large, heavy cars have a harder task meeting CAFE targets than those who specialize in making small, light cars. These disadvantages of the current CAFE system motivate the search for an alternative.

THE RELATIONSHIP BETWEEN WEIGHT AND FUEL CONSUMPTION

Figure 5A-2 fits a regression line through the car data—the upward-sloping straight line. There is strong relationship between weight and fuel consumption.[1] Figure 5A-3 adds in the data for light-duty trucks (gasoline engines only; models that sold fewer than 1,000 vehicles per year in the United States are not shown). Again, there is a strong relationship between weight and fuel consumption, though with somewhat more outliers than in the car graphs. A regression through the truck data was computed and is shown as a dashed line. It is nearly parallel to the car line.

[1] Why do the points in Figure 5A-2 scatter? Imagine a vertical line drawn at the 3,000-lb point on the weight axis, and consider the cars that fall along that 3,000-lb line. The cars do not all have the same fuel consumption because they do not all have the same powertrain technology, aerodynamic efficiency, and rolling resistance. The point where the 3,000-lb line crosses the sloping line represents the average technology of 3,000-lb cars. The sloping line is derived from a sales-weighted regression fit and therefore gives more importance to high-volume vehicles. For this reason it puts most weight on the most often used technologies within each weight class.

If all manufacturers exactly met the weight-based targets shown by the two regression lines, the total car fleet would average 28.1 mpg, and the total truck fleet would average 20.1 mpg, a difference of 8 mpg. But the two regression lines in Figure 5A-3 are only about 2.5 mpg apart. The reason for this apparent difference (8 mpg instead of 2.5 mpg) is that the regression lines estimate fuel consumption while holding weight constant. The 8-mpg car–truck gap occurs because the average car is being produced on the left-hand, low-weight end of the technology curve, while the average truck is being produced on the high weight end. Analyzing the components of the 8-mpg gap: 2.5 mpg of the gap is technological—trucks have more aerodynamic drag, and in general their drive trains are not as technologically advanced. And 5.5 mpg of the gap occurs because trucks are designed to be heavier than cars.

To gain a better sense of the characteristics of specific vehicles, Figure 5A-4 shows a sample of 33 trucks and 44 cars that are representative of cars, vans, SUVs, and pickup trucks. For analytic purposes it is sometimes more convenient to normalize the vertical scale of Figure 5A-4: Divide the fuel consumption of each point by the curb weight. This ratio is the weight-specific fuel consumption (WSFC).

Figures 5A-3, 5A-4, and 5A-5 show that fuel consumption is roughly proportional to the weight of the vehicles or, equivalently, that the weight-specific fuel consumption is roughly constant across the various weights. That is, the most significant variable explaining fuel consumption is weight. This suggests the possibility of basing fuel economy standards on the weight of the vehicle. For example, use the

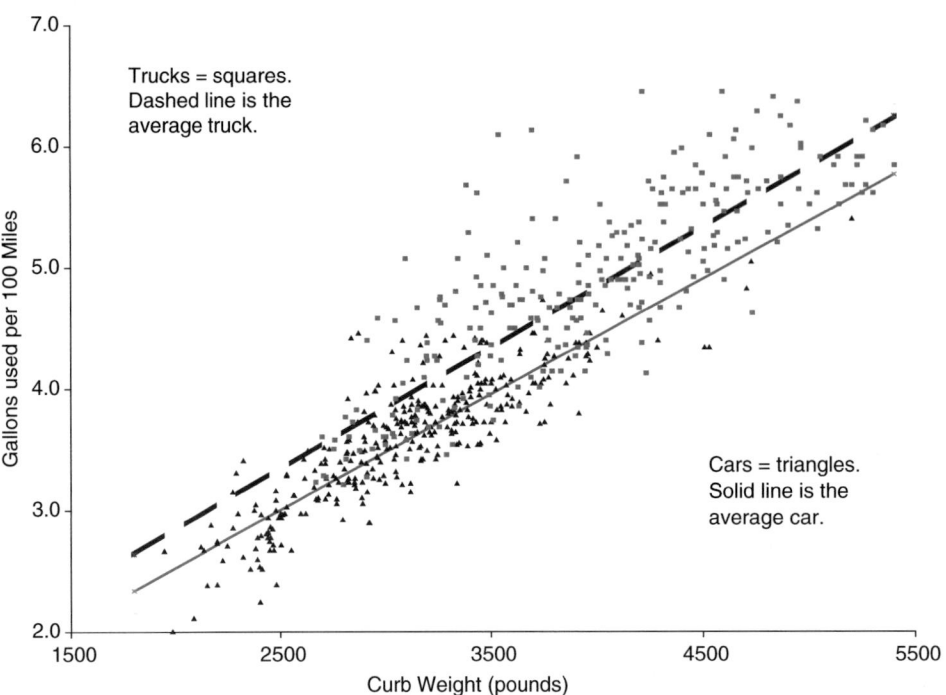

FIGURE 5A-3 Gallons to drive 100 miles, with regression lines (cars and trucks, gasoline engines only).

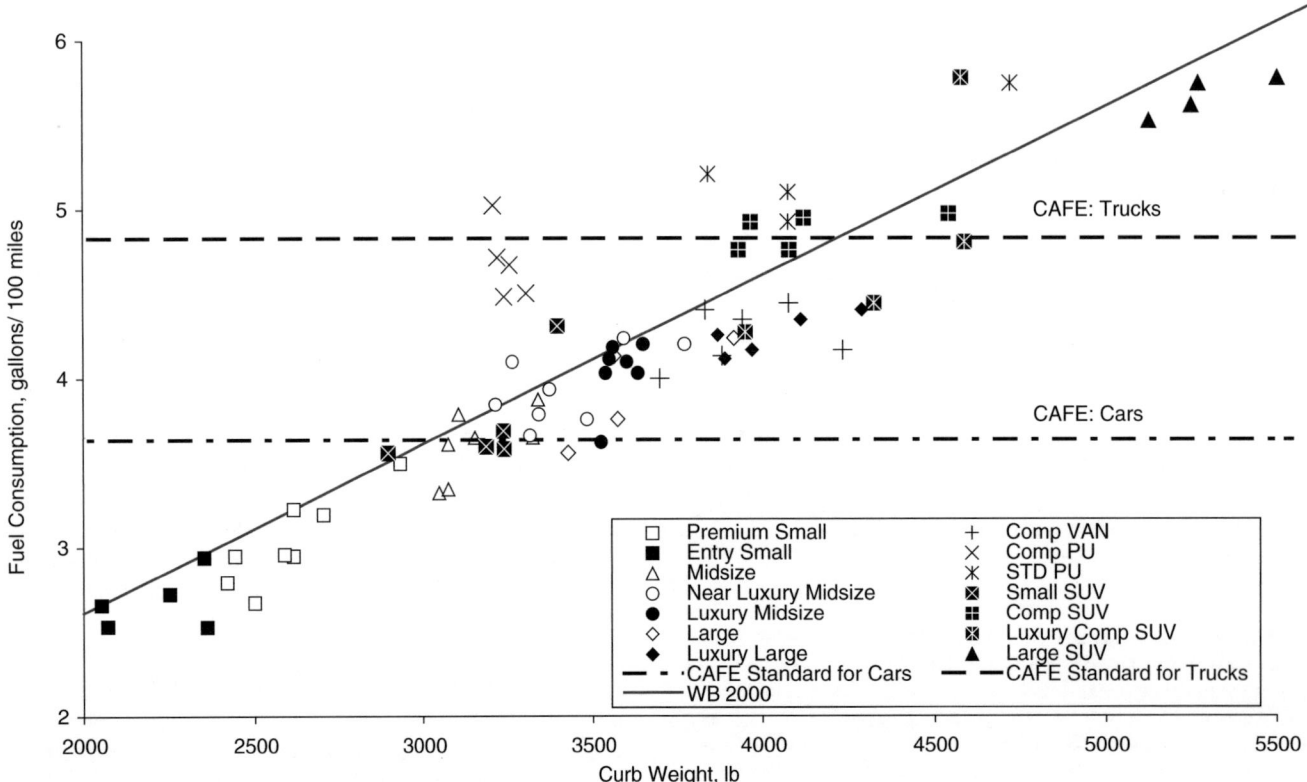

FIGURE 5A-4 Gallons used per 100 miles (all vehicles).

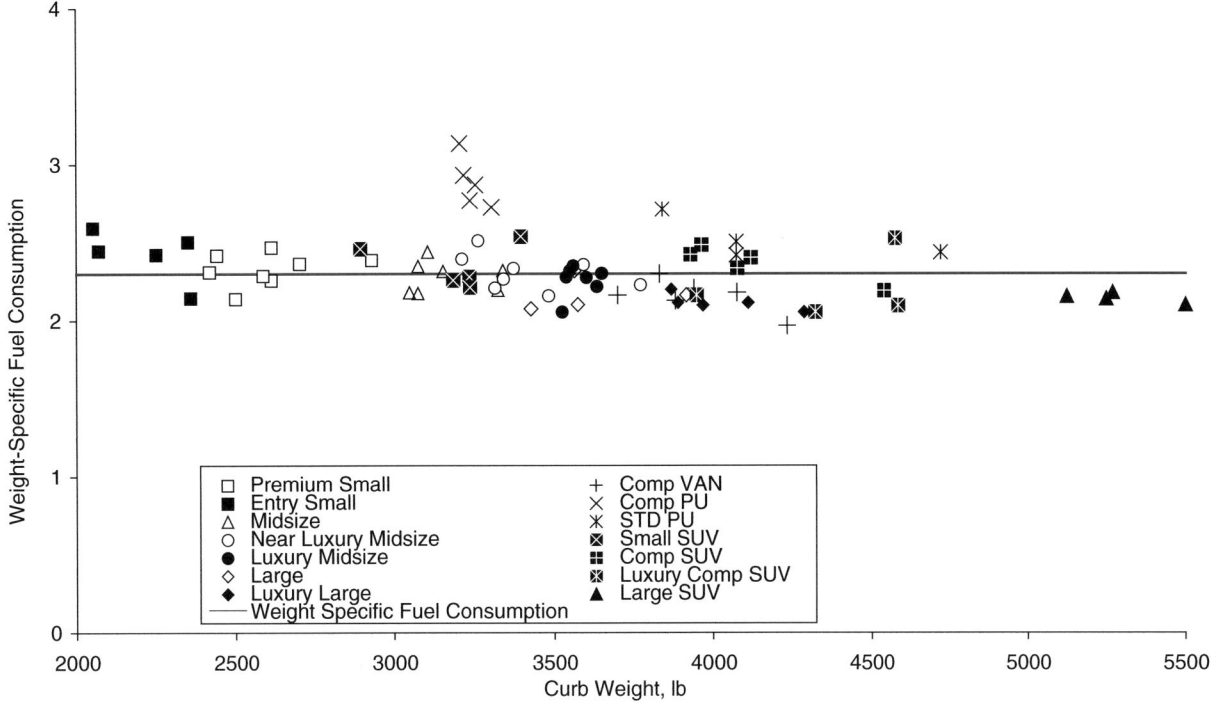

FIGURE 5A-5 Weight-specific fuel consumption (gal/100 miles/ton).

sloped line in Figure 5A-2 as the target baseline. This contrasts to the current CAFE system, which computes the gap between each vehicle and the horizontal CAFE line in Figure 5A-1. Instead, use the sloped line in Figure 5A-2 and compute the gap between each vehicle and the sloped line.

WEIGHT-BASED TARGETS VERSUS CURRENT CAFE TARGETS

A regulatory system using weight-based targets would remove the intense incentives for manufacturers to downweight their small cars, thereby reducing the potential negative safety effects of the current system. It would also produce greater equity across manufacturers—under CAFE, manufacturers who make a full range of car sizes have a harder time meeting the standards.

Weight-based targets also have three major disadvantages. First, because they are weight-neutral, the principal lever for influencing vehicle fuel economy is lost. Second, they remove most of the incentive behind the current research programs that are pursuing the use of lightweight materials to substitute for the steel in vehicles. Such programs have the potential to reduce vehicle weight while preserving vehicle size, reducing fuel consumption while preserving safety.

Third, and most important, weight-based standards could result in *higher* fuel consumption. Unlike with CAFE, there is no cap on the fleet average, so the average vehicle could move to the right (upweight) on the curve. Is this likely? Note that car weights and truck weights have been increasing over the past decade despite strong counteracting pressure from CAFE. Furthermore, the profit margin associated with large vehicles has traditionally been much higher than that associated with small ones. Thus there are substantial market incentives for manufacturers to increase vehicle weights and no restraints on their doing so once CAFE is removed.

With these advantages and disadvantages in mind, the weight-based standard could be modified to become the Enhanced-CAFE standard discussed in the next section. The committee recommends that serious consideration be given to this alternative as a substitute for the current CAFE system.

PRINCIPAL ALTERNATIVE: THE ENHANCED-CAFE STANDARD

It is possible to combine the CAFE system with weight-based targets to preserve most advantages of each while eliminating most disadvantages. In particular, the combined system should improve safety, so it is called the Enhanced-CAFE (E-CAFE) system. The E-CAFE system is a way to restructure the current regulatory system. It creates a different kind of baseline for measuring compliance and hence creates different incentives for manufacturers—incentives that move the regulatory system toward some highly desirable goals. One possible set of targets is illustrated below, but the actual targets would be determined by the legislative and regulatory process.

It is possible to have separate E-CAFE baselines for cars and for trucks. However, there would be a substantial advantage to using a single baseline that applies to all light-duty vehicles—it would eliminate the "gaming" possibilities inherent in a two-class system with different standards for each class. The horizontal line in Figure 5A-5 shows such a fuel consumption target: a single baseline used to measure performance deviations for both cars and trucks. For vehicles that weigh less than 4,000 lb, the target is sloped upward like the weight-based targets.[2] For vehicles that weigh more than 4,000 lb, the target is a horizontal line like the current CAFE standard. This approach in effect uses a more stringent target for vehicles above 4,000 lb, creating incentives to use advanced technology to improve power-train efficiency, reduce aerodynamic and rolling resistance losses, and reduce accessory power.

E-CAFE creates a strong set of incentives to improve the fuel economy of the heaviest vehicles. Under current CAFE, if a manufacturer wishes to offset the excess fuel consumption of a large vehicle, it can do so by selling a light vehicle: The vertical gap of the large vehicle ("A") in Figure 5A-1 is offset by the vertical gap of the small vehicle ("B"). But if the baseline is changed to E-CAFE (Figure 5A-6), the small vehicle does not generate a large credit because it is on the sloped portion of the baseline and its gap is measured with respect to the slope, not with respect to the horizontal line.

For our illustrative example, the horizontal line is set at 20.7 mpg, the current CAFE standard for light-duty trucks. Each manufacturer is judged on its entire fleet of cars and trucks: Vehicles that use less fuel than the targets can balance vehicles that use more. The committee recommends that a system of tradable credits, such as that described earlier in this chapter, be made part of the regulation.

The E-CAFE targets can also be expressed in terms of the weight-specific fuel consumption (WSFC) of the vehicles, which is fuel consumption per ton of vehicle weight used in 100 miles of driving. This normalized measure is shown in Figure 5A-7.

How would this proposal affect the different manufacturers? A fleetwide compliance measure was computed for each of the Big 3 manufacturers plus Honda and Toyota to measure their position with respect to the illustrative E-CAFE targets. Compliance ranged from 3 percent below the targets to 6 percent above the targets. None of the major manufacturers begins with a large compliance deviation. It is a relatively fair starting point.

The system has a single set of targets for all vehicles. This eliminates any concerns about arbitrary truck/car distinctions and their possible manipulation, since all such distinctions would be eliminated.

There would be a small incentive for lightweight vehicles to be made heavier and a large incentive for vehicles weighing more than the cutoff weight to be made lighter. Thus, the variance in weight across the combined fleet should be lower. This reduction in weight variance would improve safety in car-to-car collisions.

The present position of the lines could serve as the initial baseline under the E-CAFE system. It produces a combined car and truck fuel economy of 24.6 mpg.[3] To improve the overall fleet fuel economy in subsequent years, the horizontal portion of the baseline would be lowered, while simultaneously reducing the slope of the lower portion of the baseline. The slope of the lower portion could also be adjusted to reflect the most cost-effective use of technology (see Chapter 4). If the E-CAFE system is adopted, there should be a phase-in period associated with the new standards: Manufacturers have already made plans based on the existing CAFE standards and must be given time to analyze the implications of the new standards and to redo their product plans.

An Alternative Attribute System

Instead of basing the E-CAFE standards on curb weight, they might be based on some measure of the vehicle's load-carrying capacity, such as gross vehicle weight (GVW). Thus, vehicles capable of carrying more load would be given more liberal fuel consumption targets. This concept has some potentially useful features, as described in Attachment 3A. For regulatory purposes, however, it would have some serious problems.

Passenger vehicles rarely travel under full-load conditions. For example, data on vehicle occupancy from the Nationwide Personal Transportation Survey show that the large seating capacity of these vehicles is typically unused. The average van carries only 2.1 people, the average SUV carries 1.7 people, the average pickup carries 1.4 people, and for "other trucks" the average was 1.1 people. All these occupancy figures need to be compared with the average automobile, which carries 1.6 people. That is, if GVW were used instead of curb weight, heavier vehicles would be allowed to consume more fuel all the time because they *might* carry a full load.

There are no measured data available that would allow

---

[2]This figure is only an illustration of one possible implementation of E-CAFE. The equation for this part of the targets is: GPM100 = –0.409 + 1.31 times weight, where GPM100 is gasoline consumption in gallons per 100 miles and weight is the vehicle weight in thousands of pounds.

[3]This was calculated by assuming that every manufacturer complied with the E-CAFE standards. Compliance is measured by computing the gap between the E-CAFE baseline and each vehicle's estimated fuel-consumption, multiplied by the number of vehicles of that model that were sold—some would be positive numbers, some negative. These would be summed over all the models made by a given manufacturer.

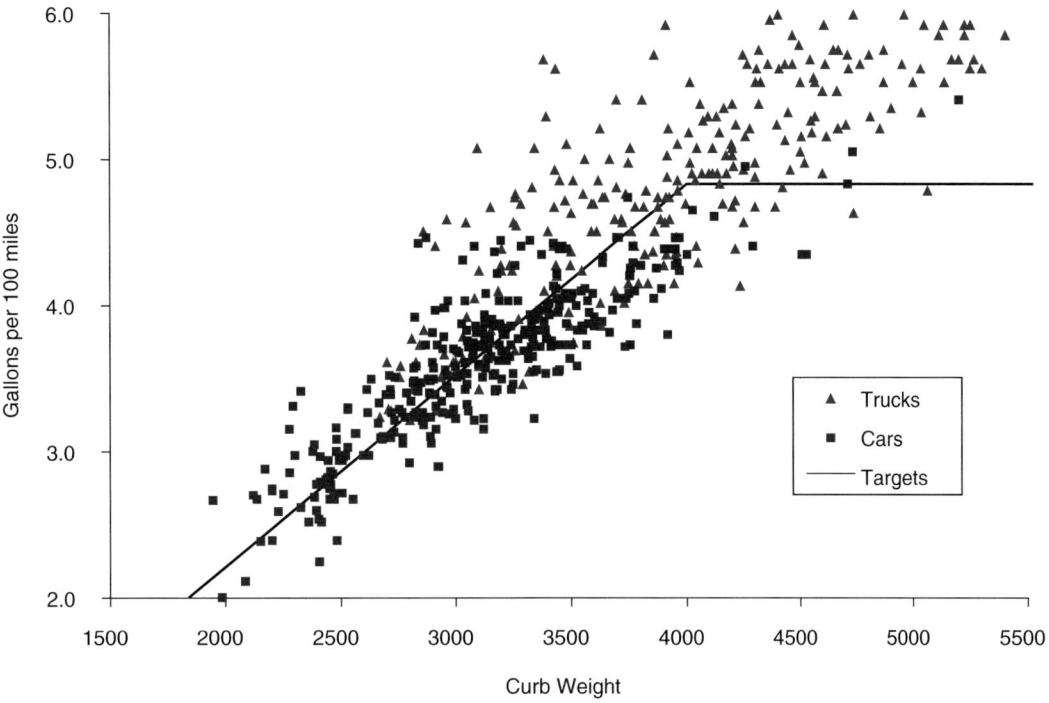

FIGURE 5A-6 Enhanced CAFE targets.

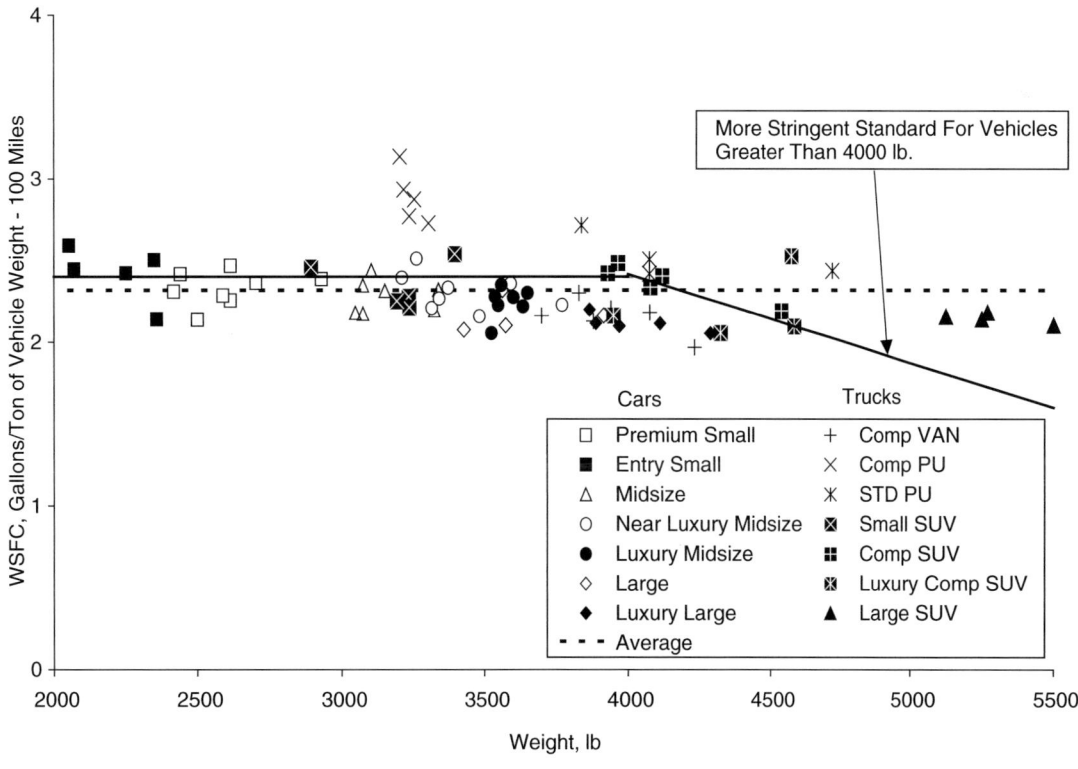

FIGURE 5A-7 Enhanced CAFE targets in WSFC units.

the determination of slopes and cutoffs under a GVW standard. EPA does not measure fuel consumption under those extra weight conditions. It has adjustment formulas capable of estimating the change in fuel consumption for small changes in vehicle weight, but the uncertainty of the predictions becomes larger and larger as the prospective weight change increases. To do an accurate analysis to set standards under the GVW criterion, the EPA would have to rerun the tests across all the vehicles in the fleet. Recalibrating to a regulatory standard based on GVW would take a long time.

Finally, GVW is a rating, not a measure, and it is determined by each manufacturer, using that manufacturer's own judgment of carrying capacity. A senior EPA analyst[4] characterized GVW as "a remarkably arbitrary figure." As currently determined, it lacks the objective reliability needed when setting a regulatory standard.

---

[4]Personal communication from Eldert Bonteko, Environmental Protection Agency, Ann Arbor, to committee member Charles Lave.

6

Findings and Recommendations

As noted in previous chapters, the committee gathered information through presentations at its open meetings (see Appendix C); invited analyses and statements; reports from its consultants, who conducted analyses at the direction of the committee; visits to manufacturers; review of the pertinent literature; and the expertise of committee members. Informed by this substantial collection of information, the committee conducted its own analyses and made judgments about the impacts and effectiveness of CAFE standards (see Chapters 2 to 5). Since Congress asked for a report by July 1, 2001, and the committee had its first meeting in early February 2001, the committee had less than 5 months (from early February to late June), to conduct its analyses and prepare a report for the National Research Council's report review process, an unusually short time for a study of such a complex issue. In its findings and recommendations, the committee has noted where analysis is limited and further study is needed.

FINDINGS

Finding 1. The CAFE program has clearly contributed to increased fuel economy of the nation's light-duty vehicle fleet during the past 22 years. During the 1970s, high fuel prices and a desire on the part of automakers to reduce costs by reducing the weight of vehicles contributed to improved fuel economy. CAFE standards reinforced that effect. Moreover, the CAFE program has been particularly effective in keeping fuel economy above the levels to which it might have fallen when real gasoline prices began their long decline in the early 1980s. Improved fuel economy has reduced dependence on imported oil, improved the nation's terms of trade, and reduced emissions of carbon dioxide, a principal greenhouse gas, relative to what they otherwise would have been. If fuel economy had not improved, gasoline consumption (and crude oil imports) would be about 2.8 million barrels per day greater than it is, or about 14 percent of today's consumption.

Finding 2. Past improvements in the overall fuel economy of the nation's light-duty vehicle fleet have entailed very real, albeit indirect, costs. In particular, all but two members of the committee concluded that the downweighting and downsizing that occurred in the late 1970s and early 1980s, some of which was due to CAFE standards, probably resulted in an additional 1,300 to 2,600 traffic fatalities in 1993.[1] In addition, the diversion of carmakers' efforts to improve fuel economy deprived new-car buyers of some amenities they clearly value, such as faster acceleration, greater carrying or towing capacity, and reliability.

Finding 3. Certain aspects of the CAFE program have not functioned as intended:

- The distinction between a car for personal use and a truck for work use/cargo transport has broken down, initially with minivans and more recently with sport utility vehicles (SUVs) and cross-over vehicles. The car/truck distinction has been stretched well beyond the original purpose.
- The committee could find no evidence that the two-fleet rule distinguishing between domestic and foreign content has had any perceptible effect on total employment in the U.S. automotive industry.
- The provision creating extra credits for multifuel vehicles has had, if any, a negative effect on fuel economy, petroleum consumption, greenhouse gas emissions, and cost. These vehicles seldom use any fuel other than gasoline yet enable automakers to increase their production of less fuel efficient vehicles.

---

[1] A dissent by committee members David Greene and Maryann Keller on the impact of downweighting and downsizing is contained in Appendix A. They believe that the level of uncertainty is much higher than stated and that the change in the fatality rate due to efforts to improve fuel economy may have been zero. Their dissent is limited to the safety issue alone.

Finding 4. In the period since 1975, manufacturers have made considerable improvements in the basic efficiency of engines, drive trains, and vehicle aerodynamics. These improvements could have been used to improve fuel economy and/or performance. Looking at the entire light-duty fleet, both cars and trucks, between 1975 and 1984, the technology improvements were concentrated on fuel economy: It improved by 62 percent without any loss of performance as measured by 0–60 mph acceleration times. By 1985, light-duty vehicles had improved enough to meet CAFE standards. Thereafter, technology improvements were concentrated principally on performance and other vehicle attributes (including improved occupant protection). Fuel economy remained essentially unchanged while vehicles became 20 percent heavier and 0–60 mph acceleration times became, on average, 25 percent faster.

Finding 5. Technologies exist that, if applied to passenger cars and light-duty trucks, would significantly reduce fuel consumption within 15 years. Auto manufacturers are already offering or introducing many of these technologies in other markets (Europe and Japan, for example), where much higher fuel prices ($4 to $5/gal) have justified their development. However, economic, regulatory, safety, and consumer-preference-related issues will influence the extent to which these technologies are applied in the United States.

Several new technologies such as advanced lean exhaust gas after-treatment systems for high-speed diesels and direct-injection gasoline engines, which are currently under development, are expected to offer even greater potential for reductions in fuel consumption. However, their development cycles as well as future regulatory requirements will influence if and when these technologies penetrate deeply into the U.S. market.

The committee conducted a detailed assessment of the technological potential for improving the fuel efficiency of 10 different classes of vehicles, ranging from subcompact and compact cars to SUVs, pickups, and minivans. In addition, it estimated the range in incremental costs to the consumer that would be attributable to the application of these engine, transmission, and vehicle-related technologies.

Chapter 3 presents the results of these analyses as curves that represent the incremental benefit in fuel consumption versus the incremental cost increase over a defined baseline vehicle technology. Projections of both incremental costs and fuel consumption benefits are very uncertain, and the actual results obtained in practice may be significantly higher or lower than shown here. Three potential development paths are chosen as examples of possible product improvement approaches, which illustrate the trade-offs auto manufacturers may consider in future efforts to improve fuel efficiency.

Assessment of currently offered product technologies suggests that light-duty trucks, including SUVs, pickups, and minivans, offer the greatest potential to reduce fuel consumption on a total-gallons-saved basis.

Finding 6. In an attempt to evaluate the economic trade-offs associated with the introduction of existing and emerging technologies to improve fuel economy, the committee conducted what it called cost-efficient analysis. That is, the committee identified packages of existing and emerging technologies that could be introduced over the next 10 to 15 years that would improve fuel economy up to the point where further increases in fuel economy would not be reimbursed by fuel savings. The size, weight, and performance characteristics of the vehicles were held constant. The technologies, fuel consumption estimates, and cost projections described in Chapter 3 were used as inputs to this cost-efficient analysis.

These cost-efficient calculations depend critically on the assumptions one makes about a variety of parameters. For the purpose of calculation, the committee assumed as follows: (1) gasoline is priced at $1.50/gal, (2) a car is driven 15,600 miles in its first year, after which miles driven declines at 4.5 percent annually, (3) on-the-road fuel economy is 15 percent less than the Environmental Protection Agency's test rating, and (4) the added weight of equipment required for future safety and emission regulations will exact a 3.5 percent fuel economy penalty.

One other assumption is required to ascertain cost-efficient technology packages—the horizon over which fuel economy gains ought to be counted. Under one view, car purchasers consider fuel economy over the entire life of a new vehicle; even if they intend to sell it after 5 years, say, they care about fuel economy because it will affect the price they will receive for their used car. Alternatively, consumers may take a shorter-term perspective, not looking beyond, say, 3 years. This latter view, of course, will affect the identification of cost-efficient packages because there will be many fewer years of fuel economy savings to offset the initial purchase price.

The full results of this analysis are presented in Chapter 4. To provide one illustration, however, consider a mid-size SUV. The current sales-weighted fleet fuel economy average for this class of vehicle is 21 mpg. If consumers consider only a 3-year payback period, fuel economy of 22.7 mpg would represent the cost-efficient level. If, on the other hand, consumers take the full 14-year average life of a vehicle as their horizon, the cost-efficient level increases to 28 mpg (with fuel savings discounted at 12 percent). The longer the consumer's planning horizon, in other words, the greater are the fuel economy savings against which to balance the higher initial costs of fuel-saving technologies.

The committee cannot emphasize strongly enough that the cost-efficient fuel economy levels identified in Tables 4-2 and 4-3 in Chapter 4 are *not* recommended fuel economy goals. Rather, they are reflections of technological possibilities, economic realities, and assumptions about parameter values and consumer behavior. Given the choice, consumers might well spend their money on other vehicle amenities, such as greater acceleration or towing capacity, rather than on the fuel economy cost-efficient technology packages.

Finding 7. There is a marked inconsistency between pressing automotive manufacturers for improved fuel economy from new vehicles on the one hand and insisting on low real gasoline prices on the other. Higher real prices for gasoline—for instance, through increased gasoline taxes—would create both a demand for fuel-efficient new vehicles and an incentive for owners of existing vehicles to drive them less.

Finding 8. The committee identified externalities of about $0.30/gal of gasoline associated with the combined impacts of fuel consumption on greenhouse gas emissions and on world oil market conditions. These externalities are not necessarily taken into account when consumers purchase new vehicles. Other analysts might produce lower or higher estimates of externalities.

Finding 9. There are significant uncertainties surrounding the societal costs and benefits of raising fuel economy standards for the light-duty fleet. These uncertainties include the cost of implementing existing technologies or developing new ones; the future price of gasoline; the nature of consumer preferences for vehicle type, performance, and other features; and the potential safety consequences of altered standards. The higher the target for average fuel economy, the greater the uncertainty about the cost of reaching that target.

Finding 10. Raising CAFE standards would reduce future fuel consumption below what it otherwise would be; however, other policies could accomplish the same end at lower cost, provide more flexibility to manufacturers, or address inequities arising from the present system. Possible alternatives that appear to the committee to be superior to the current CAFE structure include tradable credits for fuel economy improvements, feebates,[2] higher fuel taxes, standards based on vehicle attributes (for example, vehicle weight, size, or payload), or some combination of these.

Finding 11. Changing the current CAFE system to one featuring tradable fuel economy credits and a cap on the price of these credits appears to be particularly attractive. It would provide incentives for all manufacturers, including those that exceed the fuel economy targets, to continually increase fuel economy, while allowing manufacturers flexibility to meet consumer preferences. Such a system would also limit costs imposed on manufacturers and consumers if standards turn out to be more difficult to meet than expected. It would also reveal information about the costs of fuel economy improvements and thus promote better-informed policy decisions.

Finding 12. The CAFE program might be improved significantly by converting it to a system in which fuel economy targets depend on vehicle attributes. One such system would make the fuel economy target dependent on vehicle weight, with lower fuel consumption targets set for lighter vehicles and higher targets for heavier vehicles, up to some maximum weight, above which the target would be weight-independent. Such a system would create incentives to reduce the variance in vehicle weights between large and small vehicles, thus providing for overall vehicle safety. It has the potential to increase fuel economy with fewer negative effects on both safety and consumer choice. Above the maximum weight, vehicles would need additional advanced fuel economy technology to meet the targets. The committee believes that although such a change is promising, it requires more investigation than was possible in this study.

Finding 13. If an increase in fuel economy is effected by a system that encourages either downweighting or the production and sale of more small cars, some additional traffic fatalities would be expected. However, the actual effects would be uncertain, and any adverse safety impact could be minimized, or even reversed, if weight and size reductions were limited to heavier vehicles (particularly those over 4,000 lb). Larger vehicles would then be less damaging (aggressive) in crashes with all other vehicles and thus pose less risk to other drivers on the road.

Finding 14. Advanced technologies—including direct-injection, lean-burn gasoline engines; direct-injection compression-ignition (diesel) engines; and hybrid electric vehicles—have the potential to improve vehicle fuel economy by 20 to 40 percent or more, although at a significantly higher cost. However, lean-burn gasoline engines and diesel engines, the latter of which are already producing large fuel economy gains in Europe, face significant technical challenges to meet the Tier 2 emission standards established by the Environmental Protection Agency under the 1990 amendments to the Clean Air Act and California's low-emission-vehicle (LEV II) standards. The major problems are the Tier 2 emissions standards for nitrogen oxides and particulates and the requirement that emission control systems be certified for a 120,000-mile lifetime. If direct-injection gasoline and diesel engines are to be used extensively to improve light-duty vehicle fuel economy, significant technical developments concerning emissions control will have to occur or some adjustments to the Tier 2 emissions standards will have to be made. Hybrid electric vehicles face significant cost hurdles, and fuel-cell vehicles face significant technological, economic, and fueling infrastructure barriers.

---

[2]Feebates are taxes on vehicles achieving less than the average fuel economy coupled with rebates to vehicles achieving better than average fuel economy.

Finding 15. Technology changes require very long lead times to be introduced into the manufacturers' product lines. Any policy that is implemented too aggressively (that is, in too short a period of time) has the potential to adversely affect manufacturers, their suppliers, their employees, and consumers. Little can be done to improve the fuel economy of the new vehicle fleet for several years because production plans already are in place. The widespread penetration of even existing technologies will probably require 4 to 8 years. For emerging technologies that require additional research and development, this time lag can be considerably longer. In addition, considerably more time is required to replace the existing vehicle fleet (on the order of 200 million vehicles) with new, more efficient vehicles. Thus, while there would be incremental gains each year as improved vehicles enter the fleet, major changes in the transportation sector's fuel consumption will require decades.

RECOMMENDATIONS

Recommendation 1. Because of concerns about greenhouse gas emissions and the level of oil imports, it is appropriate for the federal government to ensure fuel economy levels beyond those expected to result from market forces alone. Selection of fuel economy targets will require uncertain and difficult trade-offs among environmental benefits, vehicle safety, cost, oil import dependence, and consumer preferences. The committee believes that these trade-offs rightfully reside with elected officials.

Recommendation 2. The CAFE system, or any alternative regulatory system, should include broad trading of fuel economy credits. The committee believes a trading system would be less costly than the current CAFE system; provide more flexibility and options to the automotive companies; give better information on the cost of fuel economy changes to the private sector, public interest groups, and regulators; and provide incentives to all manufacturers to improve fuel economy. Importantly, trading of fuel economy credits would allow for more ambitious fuel economy goals than exist under the current CAFE system, while simultaneously reducing the economic cost of the program.

Recommendation 3. Consideration should be given to designing and evaluating an approach with fuel economy targets that are dependent on vehicle attributes, such as vehicle weight, that inherently influence fuel use. Any such system should be designed to have minimal adverse safety consequences.

Recommendation 4. Under any system of fuel economy targets, the two-fleet rule for domestic and foreign content should be eliminated.

Recommendation 5. CAFE credits for dual-fuel vehicles should be eliminated, with a long enough lead time to limit adverse financial impacts on the automotive industry.

Recommendation 6. To promote the development of longer-range, breakthrough technologies, the government should continue to fund, in cooperation with the automotive industry, precompetitive research aimed at technologies to improve vehicle fuel economy, safety, and emissions. It is only through such breakthrough technologies that dramatic increases in fuel economy will become possible.

Recommendation 7. Because of its importance to the fuel economy debate, the relationship between fuel economy and safety should be clarified. The committee urges the National Highway Traffic Safety Administration to undertake additional research on this subject, including (but not limited to) a replication, using current field data, of its 1997 analysis of the relationship between vehicle size and fatality risk.

Appendixes

A

Dissent on Safety Issues: Fuel Economy and Highway Safety

David L. Greene
and Maryann Keller

The relationship between fuel economy and highway safety is complex, ambiguous, poorly understood, and not measurable by any known means at the present time. Improving fuel economy could be marginally harmful, beneficial, or have no impact on highway safety. The conclusions of the majority of the committee stated in Chapters 2 and 4 are overly simplistic and at least partially incorrect.

We make a point of saying fuel economy and safety rather than weight or size and safety, because fuel economy is the subject at hand. While reducing vehicle weight, all else equal, is clearly one means to increasing fuel economy, so is reducing engine power, all else equal. To the extent that consumers value power and weight, manufacturers will be reluctant to reduce either to improve mpg. Indeed, Chapter 3 of this report, which addresses the likely means for improving passenger car and light-truck fuel economy, sees very little role for weight or horsepower reduction in comparison with technological improvements. However, we will spend most of this appendix discussing the relationships between vehicle weight and safety, because the more important technological means to improving fuel economy appear to be neutral or beneficial to safety.

In analyzing the relationships between weight and safety it is all too easy to fall into one of two logical fallacies. The first results from the very intuitive, thoroughly documented (e.g., Evans, 1991, chapter 4, and many others), and theoretically predictable fact that in a collision between two vehicles of unequal weight, the occupants of the lighter vehicle are at greater risk. The fallacy lies in reasoning that, therefore, reducing the mass of all vehicles will increase risks in collisions between vehicles. This is a fallacy because it is the relative weight of the vehicles rather than their absolute weight that, in theory, leads to the adverse risk consequences for the occupants of the lighter vehicle. In fact, there is some evidence that proportionately reducing the mass of all vehicles would have a beneficial safety effect in vehicle collisions (Kahane, 1997, tables 6-7 and 6-8; Joksch et al., 1998, p. ES-2).

The second fallacy arises from failing to adequately account for confounding factors and consequently drawing conclusions from spurious correlations. In analyzing real crashes, it is generally very difficult to sort out "vehicle" effects from driver behavior and environmental conditions. Because the driver is generally a far more important determinant of crash occurrences than the vehicle and a significant factor in the outcomes, even small confounding errors can lead to seriously erroneous results. Evans (1991, pp. 92–93), for example, cites research indicating that the road user is identified as a major factor in 95 percent of traffic crashes in the United Kingdom and 94 percent in the United States. The road environment is identified as a major factor in 28 percent and 34 percent of U.K. and U.S. crashes, respectively, while the comparable numbers for the vehicle are 8 and 12 percent. Of the driver, environment, and vehicle, the vehicle is the least important factor in highway fatalities. Moreover, there are complex relations among these factors: Younger drivers tend to drive smaller cars, smaller cars are more common in urban areas, older drivers are more likely to be killed in crashes of the same severity, and so on. To isolate the effects of a less important factor from the effects of more important yet related factors is often not possible. In the case of vehicle weight and overall societal highway safety, it appears that there are not adequate measures of exposure with which to control for confounding factors so as to isolate the effects of weight alone.

THE PROBLEM IS COMPLEX

Part of the difficulty of estimating the true relationships between vehicle weight and highway safety is empirical: reality presents us with poorly designed experiments and incomplete data. For example, driver age is linearly related to vehicle weight (Joksch, personal communication, June 19, 2001), and vehicle weight, size, and engine power are all strongly correlated. This makes it difficult to disentangle driver effects from vehicle effects. As another example, pedestrian fatalities are most concentrated in dense urban areas, where smaller vehicles predominate. In Washington, D.C., 42 percent of traffic fatalities are pedestrians; in Wyoming only 3 percent are pedestrians (Evans, 1991, p. 4). Failing to accurately account for where vehicles are driven could lead one to conclude that smaller vehicles are more likely to hit pedestrians than larger vehicles. Measures of vehicle exposure with which to control for confounding influences of drivers, environment, and other vehicle characteristics are almost always inadequate. Under such circumstances it is all too easy for confounding effects to result in biased inferences.

Another part of the problem is the systematic nature of the relationships. To fully analyze the effect of weight on safety, one must consider its impacts on both the probability of a crash (crash involvement) and the consequences of a crash (crashworthiness or occupant protection). Crashes among all types of highway users must be considered—not just crashes between passenger cars, or even all light-duty vehicles, but also crashes between light vehicles and heavy trucks, pedestrians, and cyclists, as well as single-vehicle crashes.

Only one study has attempted to fully address all of these factors. That is the seminal study done by C. Kahane of the National Highway Traffic Safety Administration (Kahane, 1997). No other study includes pedestrian and cyclist fatalities. No other study also explicitly addresses crash involvement and occupant protection. Kahane's study stands alone as a comprehensive, scientific analysis of the vehicle weight and safety issue. It makes the most important contribution to our understanding of this issue that has been made to date.

But even Kahane's study has important limitations. As the author himself noted, he was unable to statistically separate the effects of vehicle size from those of vehicle weight. This would have important implications if material substitution becomes the predominant strategy for reducing vehicle weight, since material substitution allows weight to be reduced without reducing the size of vehicles. Both the steel and aluminum industries have demonstrated how material substitution can produce much lighter vehicles without reducing vehicle dimensions (e.g., see, NRC, 2000, pp. 46–49). Not only prototype but also production vehicles have confirmed the industries' claims that weight reductions of 10 to 30 percent are achievable without reducing vehicle size.

Kahane's analysis (1997) is thorough and careful. It details at length the approximations and assumptions necessitated by data limitations. These have also been enumerated in two critical reviews of the work by the NRC (North, 1996) and industry consultants (Pendelton and Hocking, 1997). We will not belabor them here. It is important, however, to repeat the first finding and conclusion of the panel of eight experts who reviewed the Kahane (1997) study, because it is identical to our view of this issue. We quote the panel's No. 1 finding and conclusion in full.

1. The NHTSA analysts' most recent estimates of vehicle weight-safety relationships address many of the deficiencies of earlier research. Large uncertainties in the estimates remain, however, that make it impossible to use this analysis to predict with a reasonable degree of precision the societal risk of vehicle downsizing or downweighting. These uncertainties are elaborated below. (North, 1996, p. 4.)

Despite these limitations, Kahane's analysis is far and away the most comprehensive and thorough analysis of this subject. We will return to it below for insights on several issues.

THE LAWS OF PHYSICS

There is no fundamental scientific reason why decreasing the mass of all highway vehicles must result in more injuries and fatalities. In debates about CAFE and safety, it has frequently been claimed that the laws of physics dictate that smaller, lighter vehicles must be less safe. This assertion is quite true from the perspective of a single private individual considering his or her own best interests and ignoring the interests of others, but it is false from a societal perspective. Therefore, the safety issues surrounding a general downweighting or downsizing of highway vehicles are concerned with the details of how vehicle designs may change, differences in the performance of lighter weight materials, the precise distribution of changes in mass and size across the fleet, and interactions with other highway users.

The One Point on Which Everyone Agrees

There is no dispute, to the best of our knowledge, that if a collision between two vehicles of different mass occurs, the occupants of the heavier vehicle will generally fare better than the occupants of the lighter vehicle. The evidence on this point is massive and conclusive, in our opinion. This conclusion is founded in the physical laws governing the changes in velocity when two objects of differing mass collide. In a direct head-on collision, the changes in velocity (Δv) experienced by two objects of differing mass are inversely proportional to the ratio of their masses (Joksch et al., 1998, p. 11), as shown in the following equations:

$$\Delta v_1 = \frac{m_2}{m_1 + m_2}(v_1 + v_2) \quad \Rightarrow \quad \frac{\Delta v_1}{\Delta v_2} = \frac{m_2}{m_1} \quad (1)$$

Because the human body is not designed to tolerate large, sudden changes in velocity, Δv, correlates extremely well with injuries and fatalities. Empirically, fatality risk increases with the fourth power of Δv (Joksch et al., 1998). The implications are extreme. If vehicle 2 weighs twice as much as vehicle 1, the fatality risks to occupants of vehicle 1 will be approximately $2^4 = 16$ times greater than those to the occupants of vehicle 2 in a head-on collision. Lighter vehicles will generally experience greater Δv's than heavier vehicles, and their occupants will suffer greater injuries as a result. Evans (1991, p. 95) has summarized this relationship in two laws:

> When a crash occurs, other factors being equal,
>
> 1. The lighter the vehicle, the less risk posed to other road users.
> 2. The heavier the vehicle, the less risk posed to its occupants.

Evans' two laws make it clear that there are winners and losers in the mass equation. In free markets, this relationship causes a kind of market failure called an externality, which leads to oversized and overweighed vehicles. This market failure, combined with the aggressive designs of many heavier vehicles, is very likely a much more important societal safety concern than improving fuel economy. It is well known that in collisions with sport utility vehicles, pick-up trucks, and vans, car drivers are at a serious safety disadvantage, not only because of the disparity in vehicle weights but because of the aggressivity of light-truck designs (Joksch, 2000).

The simple relationship expressed by equation (1) tells us two important things. First, suppose that the masses of both vehicle 1 and vehicle 2 are reduced by 10 percent. This is equivalent to multiplying both m_2 and m_1 by 0.9. The result is a canceling of effects and no change in the Δv's. *Thus, this simple application of the laws of physics would predict that a proportionate downweighting of all light vehicles would result in no increase in fatalities or injuries in two-car crashes.* We emphasize this point because it is entirely consistent with the findings of Kahane's seminal study (1997) of the effects of downsizing and downweighting on traffic fatalities. Second, the distribution of vehicle weights *is* important. Because the probability of fatalities increases at an increasing rate with Δv, a vehicle population with widely disparate weights is likely to be less safe than one with more uniform weight, at any overall average weight.

IN COLLISIONS BETWEEN VEHICLES OF THE SAME WEIGHT, IS LIGHTER OR HEAVIER BETTER?

Kahane's results (1997) suggest that in car-to-car or light truck-to-light truck collisions, if both vehicles are lighter, fatalities are reduced. The signs of the two coefficients quantifying these effects are consistent for the two vehicle types, but neither is statistically significant. Focusing on the crashworthiness and aggressivity of passenger cars and light trucks in collisions with each other, Joksch et al. (1998) studied fatal accidents from 1991 to 1994 and found stronger confirmation for the concept that more weight was, in fact, harmful to safety.

In their analysis of the effects of weight and size in passenger car and light-truck collisions, Joksch et al. (1998) paid special attention to controlling for the age of the occupants and recognizing nonlinear relationships between key variables. Their analysis led them to two potentially very important conclusions: (1) increased weight of all cars was not necessarily a good thing for overall safety and (2) greater variability of weights in the vehicle fleet was harmful.

> Among cars, weight is the critical factor. Heavier cars impose a higher fatality risk on the drivers of other cars than lighter cars. A complement to this effect is that the driver fatality risk in the heavier car is lower. However, the reduction in the fatality risk for the driver of the heavier car is less than the increase of the fatality risk for the driver of the lighter car. Thus, the variation of weight among cars results in a net increase of fatalities in collisions. (Joksch et al., 1998, p. 62.)

Studies like those of Kahane (1997) and Joksch et al. (1998) that take greater pains to account for confounding factors appear to be less likely to find that reducing weight is detrimental to highway safety in vehicle-to-vehicle crashes than studies that make little or no attempt to control for confounding factors. This suggests to us that confounding factors are present and capable of changing the direction of a study's conclusions.

FINALLY, ALL THINGS CONSIDERED (KAHANE)

The most comprehensive assessment of the impacts of vehicle weight and size on traffic safety was undertaken by the National Highway Traffic Safety Administration, partly in response to a request from the 1992 NRC study *Automotive Fuel Economy*. The NRC study pointed out that a societal perspective that included all types of crashes and all highway users was needed, and that crash involvement as well as crashworthiness needed to be considered. The 1992 NRC study also noted the possibility that the downweighting of vehicles could increase or decrease fatalities, depending on the resulting weight distribution. Kahane's study (1997) attempted to address all of these issues. Because of its thoroughness, technical merit, and comprehensiveness, it stands as the most substantial contribution to this issue to date.

Based on traffic fatality records for model years 1985 to 1993, Kahane (1997) estimated the change in fatalities attributable to 100-lb reductions in the average weight of passenger cars and light trucks. The author carefully and prominently notes that the data and model did not allow a distinction between weight and other size parameters such as track width or wheelbase. This implies that the 100-lb downweighting includes the effects of whatever downsizing is correlated with it in the fleet under study. The report also meticulously documents a number of data problems and limitations, and the procedures used to circumvent them. While these problems have important implications for interpreting the study's results (more will be said on this subject below), in our opinion, the seriousness and professionalism with which Kahane tried to address them cannot be questioned.

Perhaps the most interesting implication of Kahane's

TABLE A-1 Estimated Effects of a 10 Percent Reduction in the Weights of Passenger Cars and Light Trucks

Type of Crash	Cars			Light Trucks		
	Fatalities in 1993 Crashes	Effect (%)	Change in Fatalities	Fatalities in 1993 Crashes	Effect (%)	Change in Fatalities
Single-vehicle						
Rollover	1,754	4.58	272	1,860	0.81	67
Object	7,456	1.12	283	3,263	1.44	208
Subtotal			555			275
Crashes with others						
Pedestrian	4,206	−0.46	−66	2,217	−2.03	−199
Big truck	2,648	1.40	126	1,111	2.63	129
Car	5,025	−0.62	−105	5,751	−1.39	−354
Light truck	5,751	2.63	512	1,110	−0.54	−27
Subtotal			467			−451

Subtotal single-vehicle crashes: 555 + 275 = 830
Subtotal crashes with others: 467 − 451 = 16
Total 846 ± 147

NOTE: Weight reductions of 10 percent in MY 2000 vehicles are assumed to be 338.6 lb (0.1 × 3,386) for passenger cars and 443.2 lb (0.1 × 4,432) for light trucks. SOURCE: Based on Kahane (1997), tables 6-7 and 6-8.

(1997) study has received little attention. Following the logic of the simple laws of physics of equation (1), one would predict that if the weight of all light-duty vehicles were reduced by an equal proportion, there would be no change in fatalities in crashes among these vehicles. Calculating a 10 percent change in weight for model year 2000 passenger cars and light trucks, and applying Kahane's estimates of the percent changes in fatalities per 100 lb of weight, one sees that Kahane's model also predicts little or no change in fatalities. The calculations are shown in Table A-1. Adding the change in fatalities in car-to-car, car-to-light truck, light truck-to-light truck, and light truck-to-car collisions produces a net change of +26, a result that is not close to being statistically different from zero.[1] Of course, the simple laws of physics say nothing about crash avoidance, which Kahane's study partially addresses. Nonetheless, these results provide empirical evidence that, from a societal perspective, an appeal cannot be made to the laws of physics as a rational for the beneficial effects of weight in highway crashes. Of course, if cars are downweighted and downsized more than light trucks, the increased disparity in weights would increase fatalities. Conversely, if trucks are downsized and downweighted more than cars, the greater uniformity would reduce fatalities. These results are also entirely consistent with the conclusions of Joksch et al. (1998).

Other studies have predicted substantial increases in fatalities for just such vehicle-to-vehicle collisions (e.g., Partyka, 1989; Lund et al., 2000). But these studies make much more modest attempts to correct for confounding factors. The Kahane (1997) and Joksch et al. (1998) studies suggest that the more thoroughly and carefully one controls for confounding effects, the weaker the apparent relationships between vehicle weight and highway fatalities become. This is evidence, albeit inconclusive, that adequately correcting for confounding effects might reduce or even eliminate the correlations between weight and overall highway safety.

Kahane's study also found that downweighting and downsizing cars and light trucks would benefit smaller, lighter highway users (pedestrians and cyclists). But the benefits to pedestrians are approximately canceled by the harmful effects to light-duty-vehicle occupants in collisions with larger, heavier highway vehicles (i.e., trucks and buses). Kahane's finding that downweighting and downsizing are beneficial to pedestrians and cyclists is important because no other study includes impacts on pedestrians, a fact that biases the other studies toward finding negative safety impacts. An important result is that in downsizing and downweighting light-duty vehicles, there will be winners and losers. Including pedestrians, cyclists, and heavy truck collisions leads to an even smaller net change for collisions among all highway users of +16, again not close to being statistically different from zero.

The bottom line is that if the weights of passenger cars and light trucks are reduced proportionally, Kahane's study predicts that the net effect on highway fatalities in collisions

---
[1] The authors confirmed with Dr. Kahane that the calculations shown in Table A-1 were consistent with the proper interpretation of his model.

among all highway users is approximately zero. Given the history of the debate on this subject, this is a startling result.

The story for single-vehicle accidents, however, is not good. Kahane's model predicts that fatalities in rollovers would increase by over 300 and fatalities in fixed object crashes by almost 500, for a total of 800 more annual fatalities. These numbers are statistically significant, according to the model. Thus, the predicted increase in fatalities due to downweighting and downsizing comes entirely in single-vehicle accidents. This is puzzling because there appears to be no fundamental principle that underlies it. Rollover propensity and crashworthiness in collisions with fixed objects should, with the exception of crashes with breakable or deformable objects, be a matter of vehicle design rather than mass. This issue will be taken up next.

WHY QUESTION THE RESULTS FOR SINGLE-VEHICLE ACCIDENTS?

Kahane's results (1997) for single-vehicle accidents are suspect, though not necessarily wrong, because other objective measures of rollover stability and crashworthiness in head-on collisions with fixed objects are not correlated with vehicle weight. An advantage of crash test data and engineering measurements is that they are controlled experiments that completely separate vehicle effects from driver and environmental effects. A disadvantage is that they may oversimplify real-world conditions and may measure only part of what is critical to real-world performance. Nonetheless, the fact that such objective measures of vehicle crashworthiness and rollover potential do not correlate with vehicle weight is cause for skepticism.

In Figures A-1 and A-2 we show the National Highway Traffic Safety Administration's five-star frontal crash test results for MY 2001 passenger cars plotted against vehicle weight (NHTSA, 2001). What is clear to the naked eye is confirmed by regression analysis. There is no statistically significant relationship between mass and either driver-side or passenger-side crash test performance.

A plausible explanation for this result may lie in the fact

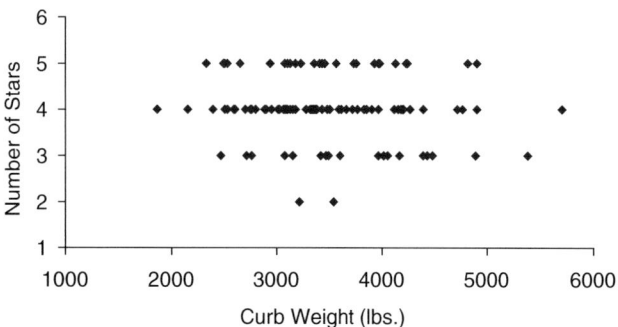

FIGURE A-2 NHTSA driver-side crash ratings for MY 2001 passenger cars.

that as mass is reduced, the amount of kinetic energy that a vehicle must absorb in a crash is proportionately reduced. Of course, the material available to absorb this energy must also be reduced and, other things equal, so would the distance over which the energy can be dissipated (the crush space). However, vehicle dimensions tend to decrease less than proportionately with vehicle mass. Wheelbase, for example, decreases with approximately the one-fourth power of mass. That is, a 10 percent decrease in mass is associated with roughly a 2.5 percent decrease in vehicle wheelbase. With mass decreasing much faster than the length of structure available to absorb kinetic energy, it may be possible to maintain fixed-object crash performance as mass is reduced. The NHTSA crash test data suggest that this has, in fact, been done.

So we are left with the question, If lighter vehicles fare as well in fixed-barrier crash tests as heavier vehicles, why should Kahane's results (1997) indicate this as one of the two key sources of increased fatalities? There are several possibilities. First, despite Kahane's best efforts, confounding of driver, environment, and vehicle factors is very likely. Second, the crash tests could be an inadequate reflection of real-world, single-vehicle crash performance. Third, the difference could, in part, be due to the ability of vehicles with greater mass to break away or deform objects.

The issue of collisions with breakable objects was investigated by Partyka (1995), who found that there was indeed a relationship between mass and the likelihood of damaging a tree or pole in a single-vehicle crash. Partyka (1995) examined 7,452 vehicle-to-object crashes in the National Accident Sampling System. Light-duty vehicles were grouped by 500-lb increments and the relationships between weight and the probability of damaging a tree or pole estimated for the 3,852 records in which a tree or pole was contacted. Partyka concluded,

> It appears that about half of vehicle-to-object crashes involved trees and poles, and about a third of these trees or poles were damaged by the impact. Damage to the tree or

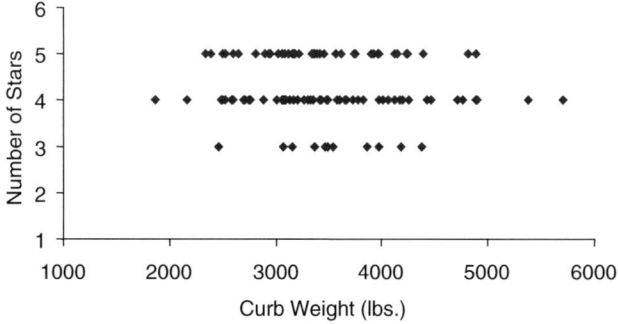

FIGURE A-1 NHTSA passenger-side crash ratings for MY 2001 passenger cars.

pole appears more likely for heavier vehicles than for lighter vehicles in front impacts, but not in side impacts.

When front and side impacts are combined, the result is an uneven relationship between mass and the probability of damaging a tree or pole, but one which generally indicates increasing probability of damage to the object with increasing vehicle mass. Figure A-3 shows the percent of time a tree or pole will be damaged by collision with passenger cars given a collision with a fixed object. Frontal and side impacts have been combined based on their relative frequency (Partyka, 1995, table 2). Roughly, the data suggest that the chances of breaking away an object may increase by 5 percent over a greater than 2,000-lb change in weight, an increase in breakaway probability of 0.25 percent per 100 lb. If one assumes that a life would be saved every time a pole or a tree was damaged owing to a marginal increase in vehicle weight (in what otherwise would have been a fatal accident), then the breakaway effect could account for about 100 fatalities per year per 10 percent decrease in light-duty vehicle weight. This is about 1 percent of annual fatalities in single-vehicle crashes with fixed objects but still represents a large number of fatalities and a potentially important concern for downweighted vehicles. While the assumption that one life would always be saved if a tree or pole broke away is probably extreme, on the other hand objects other than trees and poles can be moved or deformed.

But Partyka's study (1995) is also incomplete in that it does not address crash avoidance. To the extent that smaller and lighter vehicles may be better able to avoid fixed objects or postpone collision until their speed is reduced, there could conceivably be crash avoidance benefits to offset the reduced ability to break away or deform fixed objects. The net result is not known.

ARE LIGHTER CARS MORE LIKELY TO ROLL OVER?

The other large source of single-vehicle fatalities based on Kahane's analysis (1997) is rollover crashes. Others have also found that rollover propensity is empirically related to vehicle mass (e.g., GAO, 1994; Farmer and Lund, 2000). It is tempting to attribute these empirical results to an inherent stability conferred by mass. But there is good reason to doubt such an inference.

The stability of vehicles depends on their dimensions, especially track width, and the height of their center of gravity. If downweighting and downsizing imply a reduction in track width or a raising of a vehicle's center of gravity, the result would be greater instability. Data provided by NHTSA on its measurements of the Static Stability Factor (SSF)[2] of MY 2001 passenger cars and light trucks indicates that there is no relationship between SSF and vehicle weight *within* the car and truck classes (Figure A-4) (data supplied by G.J. Soodoo, Vehicle Dynamics Division, NHTSA, 2001). However, combining passenger cars and light trucks, one sees stability decreasing as vehicle weight increases. This is entirely due to the lesser stability of light trucks as a class.

One clear inference is that a vehicle's rollover stability based solely on the SSF is a matter of design and not inherent in its weight. Certainly, real-world performance may be far more complicated than can be captured even by theoretically valid and empirically verified measures of stability. On the other hand, the difficulty of sorting out confounding influences may also be biasing the results of statistical analyses based on real-world crashes. What raises doubts is the fact that a theoretically valid measure of vehicle stability shows no relation (or a negative relation if trucks are included) to vehicle weight. Given this, it is reasonable to surmise that some other, uncontrolled factors may account for the apparent correlation between vehicle weight and rollover fatalities.

THE BIG PICTURE (TIME SERIES DATA)

From a cursory examination of overall trends in fatality rates and light-duty vehicle fuel economy, it appears that the two move in opposite directions: fatality rates have been going down; fuel economy has been going up (Figure A-5). But the trend of declining fatality rates antedates fuel economy standards and can be observed in nearly every country in the world. Can anything be learned by statistical analysis of aggregate national fatality and fuel economy trends? Probably not.

The question is relevant because one of the earliest and most widely cited estimates of the effects of CAFE standards on traffic fatalities comes from a study by Crandall and Graham (1989) in which they regressed highway fatality rates against the average weight of *cars* on the road and other variables. The data covered the period 1947 to 1981. CAFE standards were in effect for only the last 4 years of this period, but

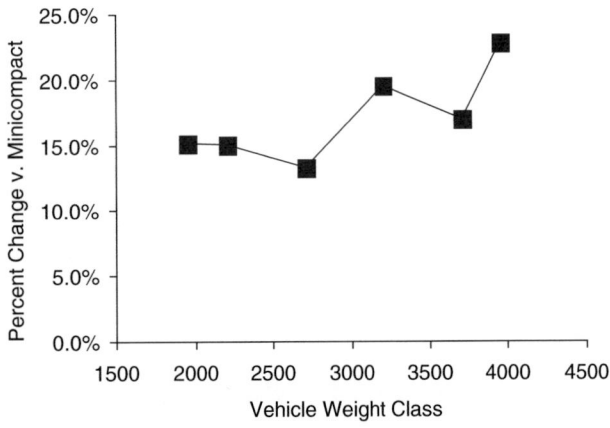

FIGURE A-3 Estimated frequency of damage to a tree or pole given a single-vehicle crash with a fixed object.

[2]NHTSA's SSF is defined as a vehicle's average track width divided by twice the height of its center of gravity. It is measured with a driver in the vehicle.

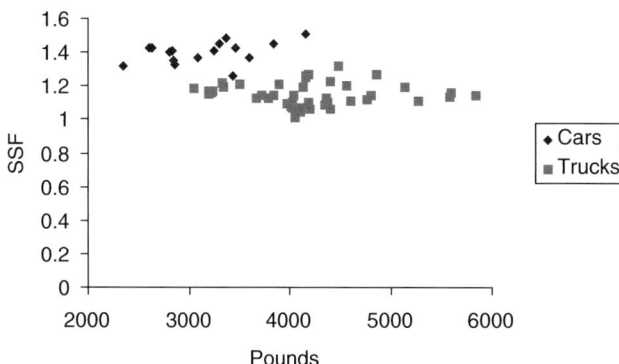

FIGURE A-4 NHTSA static stability factor vs. total weight for MY 2001 vehicles.

had been known for the last 6. Statistically significant effects of weight were found in 3 of 4 regressions presented, but other variables one might have expected to be statistically significant—including income, fraction of drivers aged 15–25, consumption of alcohol per person of drinking age, and measures of speed—were generally not significant.

Time-series regressions including variables with clear time trends, such as the declining trend of highway fatalities, are notorious for producing spurious correlations. One technique for removing such spurious correlations is to carry out the regressions on the first differences of the data. First differencing removes linear trends but retains the information produced when variables deviate from trendlines. Although it typically produces much lower correlation coefficient (R^2) values, it is generally regarded as producing more robust estimates.

We regressed total U.S. highway fatalities directly against light-duty-vehicle fuel economy and several other variables using first differences and first differences of the logarithms of the variables. The data covered the period 1966 to 1999 (data and sources available from the authors on request). No statistically significant relationship between the on-road fuel economy of passenger cars and light trucks and highway fatalities was found in any of the many model formulations we tried. Other variables tested included real GDP, total vehicle miles of highway travel, total population, the price of motor fuel, the product of the shares of light-truck and car travel, and the years in which the 55-mph speed limit was in effect. Only GDP and the 55-mph speed limit were statistically significant. The speed dummy variable assumes a value of 1 in 1975, when the 55-mph speed limit was implemented, and drops thereafter to 0.5 in 1987, when it was lifted for rural interstates, and then to 0 in 1995 and all other years. This variable may also be picking up the effects of gasoline shortages in 1974. Most often, miles per gallon appeared with a negative sign (suggesting that as fuel economy increases, fatalities decline), but always with a decidedly insignificant coefficient, as shown in typical results illustrated by the following equation:

$$\log_e(F_t) - \log_e(F_{t-1}) = -0.0458 + 1.33[\log_e(GDP_t) - \log_e(GDP_{t-1})] - 0.0875(D_t - D_{t-1}) - 0.112[\log_e(mpg_t) - \log_e(mpg_{t-1})] \quad \text{Adj. } R^2 = 0.57$$

All variables except mpg are significant at the 0.01 level based on a two-tailed t-test. The P-value for mpg is 0.69, indicating that the odds of obtaining such a result if the true relationship is zero are better than two in three. The constant suggests that fatalities would decline at 4.6 percent per year if GDP (i.e., the size of the economy) were not growing.

We present these results here only because they demonstrate that the aggregated national data covering the entire time in which fuel economy standards have been in effect and a decade before show not the slightest hint of a statistically significant relationship between light-duty-vehicle fuel economy and highway traffic fatalities. The idea that a clear and robust relationship can be inferred from aggregated national data is not supportable.

SUMMARY

The relationships between vehicle weight and safety are complex and not measurable with any reasonable degree of certainty at present. The relationship of fuel economy to safety is even more tenuous. But this does not mean there is no reason for concern. Significant fuel economy improvements will require major changes in vehicle design. Safety is always an issue whenever vehicles must be redesigned.

In addition, the distribution of vehicle weights is an important safety issue. Safety benefits should be possible if the weight distribution of light-duty vehicles could be made more uniform, and economic gains might result from even partly correcting the negative externality that encourages individuals to transfer safety risks to others by buying ever larger and heavier vehicles.

Finally, it appears that in certain kinds of accidents, reducing weight will increase safety risk, while in others it

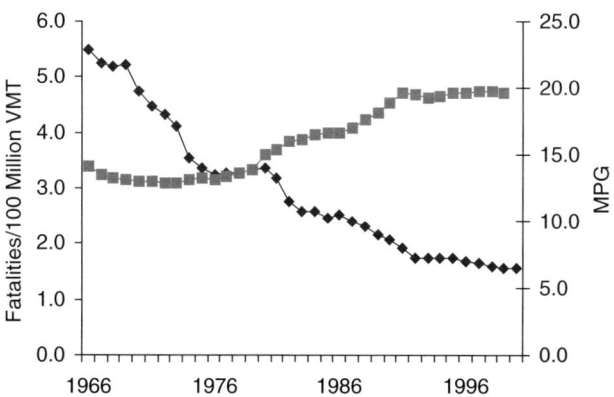

FIGURE A-5 Traffic fatality rates and on-road light-duty miles per gallon, 1996–2000.

may reduce it. Reducing the weights of light-duty vehicles will neither benefit nor harm all highway users; there will be winners and losers. All of these factors argue for caution in formulating policies, vigilance in testing vehicles and monitoring safety trends, and continued efforts to increase understanding of highway safety issues.

In conclusion, we again quote from the conclusions of the eight-member NRC panel convened to evaluate Kahane's analysis (1997) of the weight and safety issue.

> Nonetheless, the committee finds itself unable to endorse the quantitative conclusions in the reports about projected highway fatalities and injuries because of the large uncertainties associated with the results—uncertainties related both to the estimates and to the choice of the analytical model used to make the estimates. Plausible arguments exist that the total predicted fatalities and injuries could be substantially less, or possibly greater, than those predicted in the report. Moreover, possible model misspecification increases the range of uncertainty around the estimates. Although confidence intervals could be estimated and sensitivity analyses conducted to provide a better handle on the robustness of the results, the complexity of the procedures used in the analysis, the ad hoc adjustments to overcome data limitations, and model-related uncertainties are likely to preclude a precise quantitative assessment of the range of uncertainty. (North, 1996, p. 7.)

Although Kahane (1997) did estimate confidence intervals and did partially address some of the other issues raised by the NRC committee, it was not possible to overcome the inherent limitations of the data that real-world experience presented. The NRC committee's fundamental observations remain as valid today as they were in 1996.

REFERENCES AND BIBLIOGRAPHY

Crandall, Robert W., and John D. Graham. (1989). "The Effect of Fuel Economy Standards on Automobile Safety." *Journal of Law & Economics* XXXII (April).

Evans, Leonard. 1991. Traffic Safety and the Driver. New York, N.Y.: Van Nostrand Reinhold.

Evans, Leonard. 1994. "Small Cars, Big Cars: What Is the Safety Difference?" *Chance* 7 (3): 39–16.

Farmer, Charles M., and Adrian K. Lund. 2000. Characteristics of Crashes Involving Motor Vehicle Rollover, September. Arlington, Va.: Insurance Institute for Highway Safety.

GAO (General Accounting Office). 1991. Highway Safety: Have Automobile Weight Reductions Increased Highway Fatalities? GAO/PEMD-92-1, October. Washington, D.C.: GAO.

GAO. 1994. Highway Safety: Factors Affecting Involvement in Vehicle Crashes. GAO/PEMD-95-3, October. Washington, D.C.: GAO.

GAO. 2000. Automobile Fuel Economy: Potential Effects of Increasing the Corporate Average Fuel Economy Standards. GAO/RCED-00-194, August. Washington, D.C.: GAO.

Joksch, Hans C. 1985. Small Car Accident Involvement Study. Draft final report prepared for DOT, October. Washington, D.C.: NHTSA.

Joksch, Hans C. 2000. Vehicle Design Versus Aggressivity. DOT HS 809 194. April. Washington, D.C.: NHTSA.

Joksch, Hans, Dawn Massie, and Robert Pichler. 1998. Vehicle Aggressivity: Fleet Characterization Using Traffic Collision Data. DOT HS 808 679, February. Washington, D.C.: NHTSA.

Kahane, Charles J. 1997. Relationships Between Vehicle Size and Fatality Risk in Model Year 1985-93 Passenger Cars and Light Trucks. DOT HS 808 570, January. Washington, D.C.: NHTSA.

Khazzoom, J. Daniel. (1994). "Fuel Efficiency and Automobile Safety: Single-Vehicle Highway Fatalities for Passenger Cars." *The Energy Journal* 15(4): 49–101.

Lund, Adrian K., Brian O'Neill, Joseph M. Nolan, and Janella F. Chapline. 2000. Crash Compatibility Issue in Perspective. SAE Technical Paper Series, 2000-01-1378, SAE International, March. Warrendale, Pa.: SAE.

NHTSA (National Highway Traffic Safety Administration). 1995. Impacts with Yielding Fixed Objects by Vehicle Weight, NHTSA Technical Report, DOT HS 808 574, June. Washington, D.C.: NHTSA.

NHTSA. 1996. Effect of Vehicle Weight on Crash-Level Driver Injury Rates, NHTSA Technical Report, DOT HS 808 571, December. Washington, D.C.: NHTSA.

NHTSA. 1997. Relationships between Vehicle Size and Fatality Risk in Model Year 1985-1993 Passenger Cars and Light Trucks. NHTSA Technical Report, DOT HS 808 570, January. Washington, D.C.: NHTSA.

NHTSA. 2001. New Car Assessment Program. U.S. Department of Transportation (DOT). Available online at <www.nhtsa.dot.gov/NCAP>.

North, D. Warner. 1996. National Research Council, Transportation Research Board, Letter to Ricardo Martinez, National Highway Traffic Safety Administration, Department of Transportation, June 12.

NRC (National Research Council). 1992. Automotive Fuel Economy. Washington, D.C.: National Academy Press.

NRC. 2000. Review of the Research Program of the Partnership for a New Generation of Vehicles. Sixth Report. Washington, D.C.: National Academy Press.

Partyka, Susan C. 1989. Registration-based Fatality Rates by Car Size from 1978 through 1987, Papers on Car Size—Safety and Trends, NHTSA Technical Report, DOT HS 807 444, June. Washington, D.C.: NHTSA.

Partyka, Susan C. 1995. Impacts with Yielding Fixed Objects by Vehicle Weight, NHTSA Technical Report, DOT HS 808-574, June. Washington, D.C.: NHTSA.

Pendleton, Olga, and Ronald R. Hocking. 1997. A Review and Assessment of NHTSA's Vehicle Size and Weight Safety Studies, October. Ishpeming, Mich.: Pen-Hock Statistical Consultants.

B

Biographical Sketches of Committee Members

Paul R. Portney, chair, is president of and senior fellow at Resources for the Future. His former positions at RFF include vice president and senior fellow; director, Center for Risk Management; and director, Quality of the Environment Division. His previous positions include visiting lecturer, Princeton University; senior staff economist, Council on Environmental Quality; and visiting professor, Graduate School of Public Policy, University of California, Berkeley. He has served on a number of National Research Council committees, including the Committee on Epidemiology of Air Pollutants, the Board on Environmental Sciences and Toxicology, the National Forum on Science and Technology Goals, and the Committee on Opportunities in Applied Environmental Research and Development: Environment. He is the former chair of the Environmental Economics Advisory Committee of the Environmental Protection Agency and of the Science Advisory Board and was an associate editor of the *Journal of Policy Analysis and Management*. He has published widely in the areas of natural resources and public policy, environmental and resource economics, applied welfare economics, and on economic and public policy issues. He has a Ph.D. in economics from Northwestern University and a B.A. in economics and mathematics from Alma College.

David L. Morrison, vice chair, is retired director of the Office of Nuclear Regulatory Research, U.S. Nuclear Regulatory Commission. His previous positions include technical director of the Energy, Resource and Environmental Systems Division, MITRE Corporation; president of the IIT Research Institute; and director of program development and management, Battelle Memorial Institute. He has been a member of the NRC's Energy Engineering Board and the National Materials Advisory Board, chaired the NRC Committee on Alternative Energy R&D Strategies, chaired the NRC Committee on Industrial Energy Conservation, and has served on a number of other NRC committees, including the Committee on Fuel Economy of Automobiles and Light Trucks and the Committee to Review the United States Advanced Battery Consortium's Electric Vehicle R&D Project Selection Process, and as chairman of the Committee to Review the R&D Strategy for Biomass-Derived Ethanol and Biodiesel Transportation Fuels. His areas of expertise include research management, energy and environmental research, materials, nuclear technology, and physical chemistry, and he has extensive experience in the assessment of energy technologies. Dr. Morrison has a B.S. degree from Grove City College and a Ph.D. in chemistry from the Carnegie Institute of Technology.

Michael M. Finkelstein is principal, Michael Finkelstein & Associates. He has held his present position since 1991. Earlier, he held a number of positions with the Department of Transportation's National Highway Traffic Safety Administration, including those of policy advisor for the Intelligent Vehicle Highway System; associate administrator for R&D; associate administrator for Rulemaking; and associate administrator for Planning and Evaluation. He has served on several NRC committees, including the Committee on Transportation Safety Management, the Committee for Review of the Automated Highway System Consortium, and the Committee for Review of the Intelligent Vehicle Initiative. He has a B.A. from Columbia University and an M.A. from Rutgers University.

David L. Greene is corporate fellow of Oak Ridge National Laboratory (ORNL). He has spent more than 20 years researching transportation and energy policy issues, including energy demand modeling, economic analysis of petroleum dependence, modeling market responses to advanced transportation technologies and alternative fuels, economic analysis of policies to mitigate greenhouse gas emissions from transportation, and developing theory and methods for measuring the sustainability of transportation systems. Dr. Greene received a B.A. degree from Columbia University in 1971, an M.A. from the University of Oregon in 1973, and a

Ph.D. in geography and environmental engineering from the Johns Hopkins University in 1978. Dr. Greene spent 1988 and 1989 in Washington, D.C., as a senior research analyst in the Office of Domestic and International Energy Policy, Department of Energy (DOE). He has published over 150 articles in professional journals, contributions to books, and technical reports. From 1997 to 2000, Dr. Greene served as the first editor in chief of the *Journal of Transportation and Statistics*, and he currently serves on the editorial boards of *Transportation Research D*, *Energy Policy*, *Transportation Quarterly*, and the *Journal of Transportation and Statistics*. Active in the Transportation Research Board (TRB) and the National Research Council, Dr. Greene has served on several standing and ad hoc committees. He is past chair and member emeritus of the TRB's Energy Committee, past chair of the Section on Environmental and Energy Concerns, and a recipient of the TRB's Pyke Johnson Award.

John H. Johnson is Presidential Professor, Department of Mechanical Engineering-Engineering Mechanics, Michigan Technological University (MTU) and a fellow of the Society of Automotive Engineers. His experience spans a wide range of analysis and experimental work related to advanced engine concepts, emissions studies, fuel systems, and engine simulation. He was previously project engineer, U.S. Army Tank Automotive Center, and chief engineer, Applied Engine Research, International Harvester Co., before joining the MTU mechanical engineering faculty. He served as chairman of the MTU mechanical engineering and engineering mechanics department from 1986 until 1993. He has served on many committees related to engine technology, engine emissions, and health effects, including committees of the Society of Automotive Engineers, the National Research Council, the Combustion Institute, the Health Effects Institute, and the Environmental Protection Agency, and he consults for a number of government and private sector institutions. In particular, he served on the NRC Committee on Fuel Economy of Automobiles and Light Trucks and its Committee on Advanced Automotive Technologies Plan and was chair of the Committee on Review of DOE's Office of Heavy Vehicle Technologies. He received his Ph.D. in mechanical engineering from the University of Wisconsin.

Maryann N. Keller was the president of priceline.com's Auto Services division until November 2000. Before that, Ms. Keller spent nearly 28 years on Wall Street, including positions at Kidder Peabody & Co., Paine Webber, and Vilas-Fischer Associates, and was the managing director of Furman Selz Incorporated. She was named one of the top three auto analysts on Wall Street for 12 consecutive years. Ms. Keller's monthly column has appeared in *Automotive Industries* magazine for more than 10 years. She has also appeared on numerous television and radio programs in the United States, Europe, and Japan and has been a frequent guest on business and financial programs aired on Bloomberg, CNN, and CNBC. Her first book, *Rude Awakening: The Rise, Fall and Struggle to Recover at General Motors*, was published in 1989 by William Morrow. Columbia University awarded her the prestigious Eccles prize for her book. Her second book, *Collision: GM, Toyota and Volkswagen and the Race to Own the Twenty-First Century*, was published by Doubleday in 1993. She has a B.S. in chemistry from Rutgers University and an MBA in finance from the Bernard Baruch School, City University of New York.

Charles A. Lave is professor of economics (emeritus) at the University of California, Irvine, and associate director of the Institute of Transportation Studies. He received a Ph.D. in economics from Stanford University. He has been a visiting scholar at Berkeley, Stanford, and the Massachusetts Institute of Technology. He was chair of the department of economics at the University of California, Irvine, for three terms. He was the American Economics Association's representative on the Board of Directors of the National Bureau of Economic Research. He has been a member of numerous Transportation Research Board standing committees, study committees, and panels. He has served on the editorial boards of four transportation journals. He received the Transportation Research Board's Pyke Johnson Award for the outstanding research paper at its 65th Annual Meeting. He received the Extraordinarius Award of the University of California, Irvine, for distinguished contributions to teaching, service, and scholarship. His research concentrates on the demand for automobile travel, the future of mass transit, and the conservation of transportation energy.

Adrian K. Lund is currently the chief operating officer for the Insurance Institute for Highway Safety (IIHS), where he has overall responsibility for research programs at the Institute and its affiliate, the Highway Loss Data Institute. Before coming to IIHS in 1981 as a psychologist and behavioral scientist, he was an assistant professor in the Department of Behavioral Sciences and Community Health at the University of Connecticut School of Dental Medicine. Since joining the institute, Dr. Lund has participated in and directed research spanning the range of driver, vehicle, and roadway factors involved in the safety of motor vehicle travel. He is a member of the Society of Automotive Engineers, the American Public Health Association, and the American Psychological Association. He also currently serves as chairman of the Side Airbag Out-of-Position Injury Technical Working Group, an effort cosponsored by the Alliance of Automobile Manufacturers, the Association of International Automobile Manufacturers, the Automotive Occupant Restraints Council, and the Insurance Institute for Highway Safety.

Phillip S. Myers (NAE) is emeritus distinguished research professor and former chairman of the Department of Mechanical Engineering, University of Wisconsin, Madison, and a fellow of the American Society of Mechanical Engineers, the Society of Automotive Engineers (SAE), and the American Association for the Advancement of Science (AAAS). He was the president in 1969 of the SAE and has served on numerous NRC committees, including the Committee on Fuel Economy of Automobiles and Light Trucks, the Committee on Toxicological and Performance Aspects of Oxygenated Motor Vehicle Fuels, the Committee on Advanced Automotive Technologies Plan, and the Committee on Review of the DOE's Office of Heavy Vehicle Technologies. He is a fellow of the SAE and the AAAS. He is a member of the National Academy of Engineering. His research interests include internal combustion engines, combustion processes, engine emissions, and fuels. He has a Ph.D. in mechanical engineering from the University of Wisconsin, Madison.

Gary W. Rogers is president, CEO, and sole director of FEV Engine Technology, Inc. He is also vice president of North American Operations, FEV Motorentechnik GmbH. Earlier, he was director, Power Plant Engineering Services Division, and senior analytical engineer, Failure Analysis Associates, Inc.; design development engineer, Garrett Turbine Engine Company; and exploration geophysicist, Shell Oil Company. He has extensive experience in research, design, and development of advanced engine and powertrain systems, including homogeneous and direct-injection gasoline engines, high-speed direct-injection (HSDI) passenger car diesel engines, heavy-duty diesel engines, hybrid vehicle systems, and gas turbines, pumps, and compressors. He provides corporate leadership for a multinational research, design, and development organization specializing in engines and energy systems. He is a member of the Advanced Powerplant Committee and the Society of Automotive Engineers and is an advisor to the Defense Advanced Research Projects Agency program on heavy-fuel engines and to the College of Engineering and Computer Science, Oakland University, Rochester, Michigan. He recently served as a member of the NRC's Committee on Review of DOE's Office of Heavy Vehicle Technologies Program. He has a B.S.M.E. from Northern Arizona University.

Philip R. Sharp is lecturer in Public Policy at the John F. Kennedy School of Government at Harvard University. He is the former director of the Institute of Politics at Harvard University and an associate of the Harvard Electricity Policy Group. He is a member of the Board of Directors of the Energy Foundation, the Cinergy Corporation, New England Electric Power (a subsidiary of National Grid USA), and Proton Energy Systems. He also serves on the Secretary of Energy Advisory Board. Dr. Sharp graduated from Georgetown University's School of Foreign Service in 1964 and received his Ph.D. in government from Georgetown in 1974. He was a 10-term member of Congress from 1975 to 1995 from Indiana and served as chair of the Energy and Power Subcommittee of the House Committee on Energy and Commerce from 1981 to 1995.

James L. Sweeney is professor of management science and engineering, Stanford University, and senior fellow, Stanford Institute for Economic Policy Research. He has been director of the Office of Energy Systems; director of the Office of Quantitative Methods; and director of the Office of Energy Systems Modeling and Forecasting, Federal Energy Administration. At Stanford University, he has been chair, Institute of Energy Studies; Director, Center for Economic Policy Research; director, Energy Modeling Forum; chair, Department of Engineering-Economic Systems; and chair, Department of Engineering-Economic Systems and Operations Research. He has served on several NRC committees, including the Committee on the National Energy Modeling System and the Committee on the Human Dimensions of Global Change and has been a member of the Board on Energy and Environmental Systems. He is currently a member of the NRC Committee on Benefits of DOE's R&D in Energy Efficiency and Fossil Energy. He is a fellow of the California Council on Science and Technology. His research and writings address economic and policy issues important for natural resource production and use; energy markets including oil, natural gas, and electricity; environmental protection; and the use of mathematical models to analyze energy markets. He has a B.S. degree from the Massachusetts Institute of Technology and a Ph.D. in engineering-economic systems from Stanford University.

John J. Wise (NAE) is retired vice president of Research, Mobil Research and Development Corporation. He has also been vice president, R&E Planning; manager of Exploration and Production R&D; manager, Process and Products R&D; director of the Mobil Solar Energy Corporation; and director of the Mobil Foundation. He served on the Board of Directors of the Industrial Research Institute and as co-chair of the Auto/Oil Air Quality Improvement Research Program. He has also served as a member and chair of numerous NRC committees, is currently a member of the Board on Energy and Environmental Systems, and recently served as a member of the Committee on Review of DOE's Office of Heavy Vehicle Technologies. He is a member of the National Academy of Engineering. He has expertise in R&D management, petroleum refining, fuels and lubricants, the effect of fuels on vehicle emissions, and synthetic fuels. He received a B.S. in chemical engineering from Tufts University and a Ph.D. in chemistry from the Massachusetts Institute of Technology.

C

Presentations and Committee Activities

1. **Committee Meeting, National Academy of Sciences, Washington, D.C., February 2–5, 2001**

    Introductions
    Paul R. Portney, committee chair

    Expectations for study; how the current CAFE program operates
    Linda Lawson, U.S. Department of Transportation, Office of Policy
    Noble Bowie, National Highway Traffic Safety Administration

    EPA's role in CAFE; brief review of the case for reducing carbon emissions
    Chris Grundler and Jeff Alson, Environmental Protection Agency

    DOE's advanced research, including PNGV; brief review of the case for reducing petroleum imports
    Ed Wall, Phil Patterson, and Barry McNutt, Department of Energy

    Automobile company programs and perspectives
    Josephine S. Cooper, CEO, Alliance of Automobile Manufacturers
    Martin B. Zimmerman and Helen O. Petrauskas, Ford Motor Company
    Bernard Robertson, DaimlerChrysler
    Van Jolissaint, DaimlerChrysler
    Mustafa Mohatarem, General Motors

    Congressional expectations for this study
    The Honorable Dianne Feinstein, United States Senate

    Technologies and incentives for improved fuel economy
    John DeCicco, independent analyst

    Scenarios for a Clean Energy Future (five-lab study)
    Steve Plotkin, Argonne National Laboratory

    The PNGV Program
    U.S. Council for Automotive Research (USCAR)
    Ronald York, General Motors
    Steve Zimmer, DaimlerChrysler
    Mike Schwarz, Ford Motor Company

    GAO report *Automobile Fuel Economy* (2000)
    Bob White and David Lichtenfeld, General Accounting Office

    Voluntary fuel economy standards in Europe and Japan
    Steve Plotkin, Argonne National Laboratory

    The need for fuel economy standards
    Ann Mesnikoff, Sierra Club

    Consumer choice
    Diane Steed, Derek Crandall, Chris Wysocki, and Bryan Little, Coalition for Vehicle Choice

    MIT report *On the Road in 2010*
    Malcolm Weiss, Energy Laboratory, MIT

    Vehicle safety
    Charles Kahane, National Highway Traffic Safety Administration

    Safety implications of CAFE
    Sam Kazman, Competitive Enterprise Institute

    Fuel economy and safety issues
    Clarence Ditlow, Center for Auto Safety

2. **Committee Subgroup Meeting, Massachusetts Institute of Technology, Cambridge, Massachusetts, February 15–16, 2001**

APPENDIX C

3. Committee Subgroup Site Visits to General Motors, Troy, Michigan; Ford Motor Company, Dearborn, Michigan; and DaimlerChrysler, Auburn Hills, Michigan, March 1–2, 2001

4. Committee Subgroup Meeting, Stanford, California, March 7–8, 2001

5. Committee Meeting, National Academy of Sciences, Washington, D.C., March 12–14, 2001

   Automobile industry plans
   Timothy MacCarthy, President and CEO, Association of International Automobile Manufacturers
   John German, Manager, Environmental and Energy Analysis, Honda Motor Company

   New developments in vehicle technology
   Peter Peterson, Director of Marketing, U.S. Steel
   Richard Klimisch, Vice President, Automotive Aluminum Association

   Potential for high-economy vehicles
   David Friedman, Union of Concerned Scientists
   Roland Hwang, Natural Resources Defense Council

   Safety
   Leonard Evans, Science Serving Society
   Marc Ross, University of Michigan
   Ken Digges, George Washington University

   Perspectives on higher fuel economy
   Douglas Greenhaus, Director of Environment, Health & Safety, National Automobile Dealers Association (NADA)
   Ronald Blum, Economist, United Auto Workers (UAW)

6. Committee Subgroup Site Visits to Honda North America, Inc., Torrance, California, and Toyota Motor North America, Inc., Torrance, California, March 22–23, 2001

7. Site Visits to Panasonic EV Energy, Mitsubishi, and Nissan, Tokyo, Japan, March 25–31, 2001

8. Committee Subgroup Meeting, Association of International Automobile Manufacturers, Honda, and Subaru, Rosslyn, Virginia, March 29, 2001

9. Committee Subgroup Meeting, Alliance of Automobile Manufacturers, Washington, D.C., March 30, 2001

10. Site Visits to Volkswagen Audi (Ingolstadt), Daimler Chrysler (Stuttgart), Porsche (Weissach), and Siemens (Regensburg), Germany, April 3–7, 2001

11. Committee Meeting, National Academy of Sciences, Washington, D.C., April 16–17, 2001

    Technology options for improving fuel economy
    John DeCicco, consultant

12. Committee Subgroup Meeting, Stanford, California, April 25, 2001

13. Committee Subgroup Meeting, Stanford, California, April 26, 2001

14. Committee Meeting, National Academy of Sciences, Washington, D.C., May 2–4, 2001

15. Committee Subgroup Meeting, FEV Engine Technology, Inc. Auburn Hills, Michigan, May 11, 2001

16. Committee Meeting, National Academy of Sciences, Washington, D.C., May 29–30, 2001

17. Committee Meeting, National Academy of Sciences, Washington, D.C., June 14–15, 2001

18. Committee Meeting, National Academy of Sciences, Washington, D.C., July 17–18, 2001

D

Statement of Work: Effectiveness and Impact of CAFE Standards

Since CAFE standards were established 25 years ago, there have been significant changes in motor vehicle technology, globalization of the industry, the mix and characteristics of vehicle sales, production capacity, and other factors. The committee formed to carry out this study will evaluate the implications of these changes, as well as changes anticipated in the next few years, on the need for CAFE, as well as the stringency and/or structure of the CAFE program in future years. The committee shall give priority in its analytical work to relatively recent developments that have not been well analyzed in existing literature reviewing the CAFE program. Specifically, these events include economic and other impacts of current levels of oil imports, advances in technological development and commercialization, the need to reduce greenhouse gas emissions, and increased market share for light trucks, including SUVs and minivans.

The study shall examine, among other factors:

(1) The statutory criteria (economic practicability, technological feasibility, need for the U.S. to conserve energy, the classification definitions used to distinguish passenger cars from light trucks, and the effect of other regulations);
(2) The impact of CAFE standards on motor vehicle safety;
(3) Disparate impacts on the U.S. automotive sector;
(4) The effect on U.S. employment in the automotive sector;
(5) The effect on the automotive consumer; and
(6) The effect of requiring CAFE calculations for domestic and non-domestic fleets.

The committee shall examine the possibility of either positive or negative impacts, if any, in each of these areas as a result of CAFE standards. The committee shall also include in that analysis a statement of both the benefits and the costs, if any, in each of the identified areas, and shall include to the extent possible both quantifiable and non-quantifiable costs or benefits. The committee may also examine a broader range of related issues appropriate to providing the most accurate possible report. For example, in reviewing possible impacts on U.S. employment in the automotive sector, the committee shall also examine the impacts on U.S. employment in other sectors of the economy from both CAFE standards and resulting reductions in oil imports if relevant to providing a complete picture of impacts on U.S. employment.

The committee shall write a report documenting its conclusions and recommendations.

E

Acronyms and Abbreviations

AAA	American Automobile Association	GFRP	glass-fiber-reinforced polymer composites
ASM/AMT	automatic shift manual transmission	GHG	greenhouse gases
AWD	all-wheel drive	GM	General Motors Corporation
		GVW	gross vehicle weight
bbl	barrel		
Btu	British thermal unit	HEV	hybrid electric vehicle
C	carbon	ICE	internal combustion engine
CAE	computer-aided engineering	IPCC	Intergovernmental Panel on Climate Change
CAFE	corporate average fuel economy (standards)		
CARB	California Air Resources Board	ISG	integrated starter/generator
CFRP	carbon-fiber-reinforced polymer composites	IVT	intake valve throttling
CIDI	compression ignition, direct injection		
CNG	compressed natural gas	lb	pound
CO	carbon monoxide	LSFC	payload-specific fuel consumption
CO_2	carbon dioxide		
CR	compression ratio	MIT	Massachusetts Institute of Technology
CVA	camless valve actuation	mmbd	million barrels per day
CVT	continuously variable transmission	MMBtu	million British thermal units
		mpg	miles per gallon
DI	direct injection		
DOE	Department of Energy	NAFTA	North American Free Trade Agreement
DOHC	double overhead cam	NHTSA	National Highway Traffic Safety Administration
DOT	Department of Transportation		
		NO_x	nitrogen oxides
EEA	Energy and Environmental Analysis	NRC	National Research Council
EIA	Energy Information Administration		
EMF	Energy Modeling Forum	OHC	overhead cam
EPA	Environmental Protection Agency	OHV	overhead valve
EV	electric vehicle	OPEC	Organization of Petroleum Exporting Countries
FC	fuel cell		
FHWA	Federal Highway Administration	PEM	proton exchange membrane
4WD	four-wheel drive	PM	particulate matter
FTP	federal test procedure	PNGV	Partnership for a New Generation of Vehicles
		ppm	parts per million
GAO	General Accounting Office	PU	pickup (truck)
GDP	gross domestic product	PZEV	partial zero emission vehicle

R&D	research and development	UAW	United Auto Workers
		UPI	uniform percentage increase
SCR	selective catalytic reduction		
SI	spark-ignition	VCR	variable compression ratio
SOHC	single overhead cam	VMT	vehicle miles traveled
SULEV	super ultralow-emission vehicle	VVLT	variable valve lift and timing
SUV	sport utility vehicle	VVT	variable valve timing
TWC	three-way catalyst	WSFC	weight-specific fuel consumption
2WD	two-wheel drive	WTT	well-to-tank
		WTW	well-to-wheels

F

Letter Report:
Technology and Economic Analysis in the Prepublication Version of the Report
Effectiveness and Impact of Corporate Average Fuel Economy (CAFE) Standards

THE NATIONAL ACADEMIES
Advisers to the Nation on Science, Engineering, and Medicine

National Academy of Sciences
National Academy of Engineering
Institute of Medicine
National Research Council

January 14, 2002

The Honorable Jeffrey W. Runge
Administrator
National Highway Traffic Safety Administration
U.S. Department of Transportation
400 7th Street, SW, Room 5320
Washington, DC 20590

Dear Dr. Runge:

    We are pleased to deliver this letter report to you. It assesses the methodology used to estimate the potential for fuel economy gains in the National Research Council's (NRC) report *Effectiveness and Impact of Corporate Average Fuel Economy (CAFE) Standards,* which was released in prepublication form in July 2001.

    The NRC undertook this additional effort because, following the July 2001 release of the CAFE report, we heard criticism from some parties that the methodology was fundamentally flawed. This letter report confirms that the methodology was quite valid. The committee that prepared the CAFE report continued to refine its analysis, and the estimates are slightly different, but the conclusions and recommendations are essentially unchanged. We have incorporated these refinements (and letter report itself) into the final CAFE report, which we expect to deliver by about the end of January.

    We plan to release this letter report to the public at 9:00 a.m. on Wednesday, January 16. At that time, copies will also be made available to news reporters, and the report will be posted on the National Academies web site.

Dr. Jeffrey W. Runge
Administrator
DOT/National Highway Safety Administration
January 14, 2002
Page 2

    We appreciate the opportunity to have been of service. If you have any questions, please do not hesitate to call me at 202-334-3000.

Sincerely,

E. William Colglazier
Executive Officer
National Research Council/National Academy of Sciences

cc:

Kenneth R. Katz, CORT DOT/National Highway Safety Administration
Office of Planning and Consumer Programs (NPS-30)
400 7th Street, SW, Room 5320
Washington, DC 20590
202-366-0842

Technology and Economic Analysis in the Prepublication Version of the Report *Effectiveness and Impact of Corporate Average Fuel Economy (CAFE) Standards*

This letter report summarizes the reexamination of several technology issues originally presented in the prepublication report by the Committee on the Effectiveness and Impact of Corporate Average Fuel Economy (CAFE) Standards. It first explains why the reexamination was undertaken and the process for doing so. It then evaluates the methodology used in Chapter 3 of the prepublication version of the report for estimating the benefits of improved technology, corrects several minor errors, and explains the results. In doing so, it stresses the committee's desire that readers focus on averaged estimates for cumulative gains and costs instead of the upper and lower bounds, which reflect the increasing uncertainty of costs and benefits as fuel efficiency is increased. It also updates and explains the economic analysis presented in Chapter 4 of the prepublication version.

REASONS FOR THIS LETTER REPORT

At the request of the U.S. Congress, the National Research Council (NRC) released a prepublication version of its report *Effectiveness and Impact of Corporate Average Fuel Economy (CAFE) Standards* in July 2001. The committee prepared the report in less than 6 months because Congress expected to address CAFE standards in 2001 and had requested guidance on technical feasibility. During the study, President George W. Bush announced that this report would be an important factor in his energy policy, prepared under the direction of Vice President Richard Cheney.

During this initial 6-month period, the committee held a series of public meetings at which representatives of automobile manufacturers, governmental agencies, and a variety of nongovernmental organizations provided information on the issues addressed in the report. The committee also visited manufacturers and major suppliers, reviewed thousands of pages of presentation and other background material, and retained consultants to provide detailed analyses.

Following the release of the prepublication report, the automotive industry challenged some of the estimates for improved fuel economy. Representatives of the Alliance of Automobile Manufacturers (AAM), General Motors, and DaimlerChrysler told the NRC in August 2001 that, in their opinion, portions of the technical analysis in Chapter 3 were fundamentally flawed and that some of the estimates for fuel economy improvements violated the principle of conservation of energy. In particular, the industry claimed that the method used to estimate incremental improvements in fuel consumption through stepwise application of technologies did not consider system-level effects and that "double-counting" of potential reductions in energy losses had occurred, especially in upper bound estimations, which resulted in the violation of the first law of thermodynamics (conservation of energy).[1]

---

[1]The largest energy loss is due to inefficiency of the engine. The maximum efficiency of a typical current spark-ignition engine is about 35 percent. The remainder of the energy in the fuel is transferred to the atmosphere as thermal energy in the exhaust or through the cooling system. Some of the technologies discussed here raise efficiency, but in general it is difficult to significantly reduce these losses. Other technologies indirectly accomplish

In response to these concerns, especially in light of the potential impact of the report's findings and recommendations on national energy policy, the committee held a public meeting on October 5, 2001. Industry representatives and several analysts with other perspectives presented their questions and concerns about the report.[2] The presentations are available in the NRC's public access file.

In addition to the allegation of violating the principle of conservation of energy, industry raised other issues including the following:

1. Some technologies are already in widespread use, so the improvement from implementing them for a particular class of vehicle is minimal.
2. Improvements from some technologies are overstated.
3. Baseline fuel economy levels do not match Environmental Protection Agency (EPA) data.
4. Some data supplied to the committee may have been misinterpreted as based on fuel consumption rather than fuel economy, leading to an overstatement of benefits.
5. Because of these errors, the break-even analysis in Chapter 4 overestimates the benefits of raising fuel economy standards.

Feng An presented some of the results from a recent report by the American Council for an Energy Efficient Economy (ACEEE) and commented on the automotive industry's presentation. He pointed out that the ACEEE analysis, which was based on detailed energy balance simulation, predicted results similar to those in the committee's report when weight reduction was excluded. He also noted that industry's treatment of engine idle-off was inaccurate and that analysis of energy losses was a matter of engineering judgment as well as exact mathematics. He concluded that some double-counting of benefits may have occurred in the committee's most optimistic estimates. However, he argued that two other factors counter this problem. First, other technologies could reasonably have been included by the committee, especially weight reduction and hybrid-electric vehicles. Second, combining technologies can produce positive synergies,[3] which may not have been considered.

David Friedman stated, among other things, that the committee had clearly eliminated most double-counting, and, insofar as some may have occurred, the committee could have considered additional technologies to achieve the same or greater levels of fuel economy. He

---

this goal; e.g., friction reduction results in less heat transfer from the radiator. Many of the engine technologies discussed here typically are applied to reduce pumping losses (the energy required to move the air for combustion through the engine), a smaller loss but one easier to reduce. As these technologies are added, pumping losses decline, reducing the potential for the next technology. If these diminishing returns are not considered, the analysis may overpredict the reduction in pumping losses, resulting in double-counting. However, many of these technologies have secondary benefits as well, which also must also be considered. The term "system-level effects" refers to these interactions.

[2]Formal presentations were made by Greg Dana of the Alliance of Automobile Manufacturers; Feng An, a consultant working with the Energy Foundation and the American Council for an Energy Efficient Economy; and David Friedman of the Union of Concerned Scientists. Accompanying Mr. Dana were Aaron Sullivan of General Motors (who made an additional informal presentation), Tom Asmus of DaimlerChrysler, Tom Kenny of Ford, and Wolfgang Groth of Volkswagen. In addition, Barry McNutt of the Department of Energy made an informal presentation.

[3]System-level effects can be positive as well as negative. The term "synergies" is used when the benefit is greater than the sum of the individual contributions.

believed that with those technologies, even the most optimistic upper bound could be achieved. He noted that losses due to aerodynamic drag, rolling resistance, and inertia can easily be reduced more than the committee had allowed, and probably at lower cost than some of the technologies that are on the committee's list. In addition, hybrid electric vehicles (HEV) may become competitive faster than the committee had assumed, and positive synergies were not always included in its analysis.

The committee, in particular the Technology Subgroup,[4] examined the concerns expressed at the October 5 public meeting, reviewed additional materials submitted by interested parties, evaluated the potential for fundamental errors in its original analysis, and wrote this report to present its findings. This effort has been limited to the technology methodology presented in Chapter 3 of the prepublication version and the potential impact any revisions would have on the economic analysis in Chapter 4.

The review uncovered several minor computational or data entry errors in the original analysis. These are identified here and corrected in the final CAFE report, scheduled for publication in early 2002. In addition, the methodology used for estimating the fuel efficiency improvements is explained in greater detail, as is the increasing uncertainty in upper and lower bounds in the prepublication version of the report. These bounds have been eliminated in the final report and in this letter report in order to help focus the reader on the average estimations.

FINDINGS

Based on its review of the information provided to it subsequent to the July 2001 release of the prepublication version of the CAFE report, in combination with additional investigations conducted by the Technology Subgroup, the committee finds as follows:

1. The fundamental findings and recommendations presented in Chapter 6 of the CAFE report are essentially unchanged. The committee still finds that "technologies exist that, if applied to passenger cars and light-duty trucks, would significantly reduce fuel consumption within 15 years" and that "assessment of currently offered product technologies suggests that light-duty trucks, including SUVs, pickups, and minivans, offer the greatest potential to reduce fuel consumption, on a total-gallons-saved basis." The only changes to the findings and recommendations presented in the prepublication version are the references to the analyses presented in Chapters 3 and 4, which have been modified as discussed in the section "Technical Discussion," below, and Attachments A through E.

2. Baseline fuel consumption averages have been revised to reflect the latest results published by EPA for model year 1999. The technology matrixes have been modified to eliminate unlikely combinations that were erroneously carried forward in the spreadsheets (see Tables 3-1 to 3-3 in Attachment A). Calculations of incremental reductions in fuel consumption for certain vehicle classes[5] also have been corrected.

---

[4] John Johnson, Gary Rogers, Phillip Myers, and David Greene.
[5] Midsize and large cars should have used camless valve actuation instead of intake valve throttling in path 3. The benefits of variable valve timing should have been 2-3 percent (instead of 1-2 percent) and variable timing and lift should have been at 1-2 percent (instead of 3-8 percent).

These changes had a mixed effect on fuel economy estimates, but the net result is to slightly lower the averages. In addition, the upper and lower bounds in Table 3-4 and Figures 3-4 to 3-13 of the prepublication version have been removed (see Attachment A). The greatly increased uncertainty as technologies were added caused considerable confusion, and the committee decided to simplify the presentation. The economic analysis has been modified to reflect these changes and several other minor modifications, as discussed in the section "Analysis of Cost-Efficient Fuel Economy Levels," below, and shown in Attachment B. These changes, which are incorporated into the final CAFE report, had no significant impact on the overall findings and recommendations of the report because the average estimates changed so slightly.

3. The committee notes that its analysis of the incremental benefits of employing additional fuel-efficient technologies was, of necessity, based largely on engineering judgment. A detailed energy balance simulation of all the technologies in all the vehicle classes could potentially improve the accuracy of the analysis, but that task was well beyond the resources of the committee. The prepublication version of the report states, "Within the time constraints of this study, the committee used its expertise and engineering judgment, supplemented by the sources of information identified above, to derive its own estimates of the potential for fuel economy improvement . . . ." The report also notes that "the committee has applied its engineering judgment in reducing the otherwise nearly infinite variations in vehicle designs and technologies that would be available, to some characteristic examples." Moreover, as confirmed during testimony presented by AAM representatives, the committee did not have sufficient proprietary technical data to conduct highly detailed simulations. Additional explanation of this estimation process is presented in the "Technical Discussion" section, below.

4. The committee acknowledges that, although it was conservative in its estimates of potential gains attributable to individual technologies (in an attempt to account for potential double-counting), some overestimation of aggregated benefits, compared to aggressive development targets, may have occurred in paths 2 and 3 in the prepublication version. Nevertheless, the committee finds that the principle of conservation of energy was not violated. Furthermore, the committee may have underestimated some potential improvements and given insufficient consideration to system-level synergies.

The committee conducted a more detailed simulation to determine whether significant overestimations of potential benefits may have inadvertently occurred. Only one case (midsize SUVs) was considered in the time available, but this case provides a general confirmation of the methodology used in the CAFE report. This analysis (detailed in the technical discussion and in Attachment C), shows that the most optimistic upper-bound estimate in the prepublication version exceeded aggressive development targets by less than 10 percent. The same analysis suggests that if pumping losses were reduced to extremely low levels (due to unthrottled operation) and friction was reduced by 30 to 40 percent (theoretically possible but not currently feasible for production engines), fuel consumption reductions would equal the most optimistic upper-bound estimate for midsize SUVs in the prepublication version.

Therefore the committee finds that its analysis did not violate any laws of energy conservation. The committee acknowledges that the uncertainty associated with any upper boundary increases significantly as additional technologies are considered. Accordingly it does not propose them as development targets.

5. All estimates (even those involving sophisticated modeling) of the costs and benefits of new technologies are uncertain. As technologies are added, the overall uncertainty increases. The committee included a wide range of costs and benefits for each technology to account for such uncertainties. However, based upon the feedback received since the release of the prepublication version, the committee believes that the increasing level of uncertainty associated with moving up each of the three paths was not sufficiently explained in Chapter 3. Additional technical discussion and clarification are therefore included below. Furthermore, the committee finds that its methodology for determining the collective uncertainty as technologies are added has produced wide upper- and lower-bound estimates that have contributed to confusion and misinterpretation of the analysis. Chapter 4 uses a statistical technique to narrow the bounds (using the values for each technology in Chapter 3 as input), as seen in Figures 4-5 and 4-6 in Attachment B. This technique maintains an approximately constant confidence bound over the range of fuel economy. Therefore, the upper and lower bounds for improved fuel consumption and associated costs are dropped from Table 3-4 and Figures 3-4 to 3-13 (see Attachment A), and only the now slightly lower averages are retained in order to focus attention on the most probable and useful results. However, the reader is cautioned that even the averages are only estimates, not exact predictions.

CONCLUSIONS

Based on the additional information provided to the committee subsequent to the July 2001 release of the prepublication version of the CAFE report, including testimony provided at the October 5, 2001, meeting, the committee concludes as follows:

1. The committee reaffirms its approach and general results: Significant gains in fuel economy are possible with the application of new technology at corresponding increases in vehicle price. Although the committee believes that its average estimates, as presented here, provide a reasonable approximation of the fuel economy levels attainable, it endorses its statement in the prepublication version—namely, that changes to CAFE standards should not be based solely on this analysis. Finding 5 of the CAFE report states: "Three potential development paths are chosen as examples of possible product improvement approaches, which illustrate the trade-offs auto manufacturers may consider in future efforts to improve fuel efficiency." The finding also notes that "economic, regulatory, safety, and consumer-preference-related issues will influence the extent to which these technologies will be applied in the United States."

2. The fuel economy estimates include uncertainties that necessarily grow with the increasing complexity of vehicle systems as fuel economy is improved. Thus for regulatory purposes, these estimates should be augmented with additional analysis of the potential for improvements in fuel economy and, especially, their economic consequences. The development approaches that manufacturers may actually pursue over the next 15 years will depend on improvements made in current systems, price competitiveness of production-intent technologies, potential technological breakthroughs, advancements in diesel emission-control technologies, and the quest for cost reduction in hybrid technology.

    Path 3 includes emerging technologies that are not fully developed and that are, by definition, less certain. The committee also recognizes that this path includes technologies that likely have not been tested together as a system. The upper and lower bounds of the paths are even more uncertain than the average. Therefore in formulating its conclusions, the committee used the path averages.

    Full analysis of systems effects, which might be better defined by more rigorous individual vehicle simulations, could suggest fuel economy improvements that are greater or less than the average estimates made by the committee. More accurate estimates would require detailed analyses of manufacturer-proprietary technical information for individual vehicle models, engines, transmissions, calibration strategies, emissions control strategies, and other factors—information to which the committee has no access. Even if such information were provided, evaluating all possible scenarios would require a prohibitive number of simulations for the committee to pursue.

3. Based on input provided subsequent to the July 2001 release of the prepublication version, the committee concludes that additional technologies, beyond those identified in the report, may also become available within the 10-15 year horizon. The committee may have underestimated the vehicle-based (e.g., aerodynamics, rolling resistance, weight reduction) benefits that may be expected within 15 years. Prototype vehicles are now being designed and tested that achieve significantly higher fuel economy (FE) than the levels considered by the committee (see the section "Future Potential," in Attachment D). While the committee has not analyzed all of these concepts (they still must surmount a series of barriers, including cost, emissions compliance, and consumer acceptance issues), it notes that they illustrate the technical potential for greater fuel economy.

4. At the August 2001 meeting, industry representatives stated that the methodology used by the committee violated the principle of conservation of energy.[6] However, at

---

[6] The industry representatives separated the technologies according to how, in their judgment, they might reduce energy losses. They expected most technologies to contribute to reducing pumping and engine friction losses. When they added all the improvements from those technologies, the total exceeded some relative value assumed to represent the combined EPA city/highway cycle for a single vehicle example. This was the basis for the claim that

the October 2001 meeting, no detailed energy balance formulations or independent analyses were presented to support this claim. Rather, industry representatives presented their judgment-based contributions of the different technologies considered by the committee to reduce energy losses. The representatives then summed these contributions, suggesting that the committee's methodology overestimates the potential improvement and thereby violates the conservation of energy principle.

The committee has several points of contention with industry's formulation of the energy balance issue: for example, the allocation of the benefits of an integrated starter-generator (with idle-off) to pumping, friction, and transmission. While turning off the engine when power is not needed (i.e., during idle or braking) does not raise the efficiency of the engine itself, it does lower the energy required for the EPA test cycles used to measure fuel efficiency. Thus idle-off effectively results in an increase in overall fuel economy, which can be realized without violating the conservation of energy. This effect varies the relationship between engine losses and fuel consumption that has historically been considered when estimating fuel economy. Regenerative braking, although not considered in the three hypothetical paths, is another example of fuel economy improvement being essentially independent of engine efficiency.

In addition, assumptions as to primary and secondary benefits must consider varying trade-offs as many new technologies are aggregated. The committee therefore concludes that differences in engineering judgment are likely to produce significantly different approximations when projecting some 15 years into the future.

The committee agrees that achieving the most optimistic (upper bound) results of path 3 in the prepublication version of the report with the technologies identified there would require overcoming great uncertainty and technical risk. The committee did not regard the upper bound as a viable production-intent projection. It is a bound, by definition, as is the lower bound, and plausible projections lie somewhere in between. Furthermore, consumer acceptance and real-world characteristics will certainly cause actual fuel economy gains to be less than the technically feasible levels presented in this study.

5. The committee reaffirms its position in Finding 6 of the CAFE report: "The committee cannot emphasize strongly enough that the cost-efficient fuel economy levels identified in Chapter 4 are *not* recommended CAFE goals. Rather, they are reflections of technological possibilities, economic realities, and assumptions about parameters values and consumer behavior." The fuel economy estimates in Chapters 3 and 4 describe the trade-offs between fuel economy improvement and increased vehicle price. They do not incorporate the value of reducing U.S. oil consumption or greenhouse gas emissions. Nor are they based on particular views of the

---

the NRC analysis violates conservation of energy. The presentation, but not the specific charge, was repeated at the October meeting, yet the detailed propriety data behind the relative assumptions were not offered.

appropriateness of government involvement. The committee provides some discussion of these issues, but the value judgments must be left to policy makers.

TECHNICAL DISCUSSION

Methodological Issues

The state of the art in overall powertrain simulation, including gas exchange, combustion, heat loss, exhaust energy, and indicated thermodynamic efficiency, has advanced with the development of computing capacity, computational fluid dynamics, and mechanical system simulation. Automotive manufacturers, subsystem suppliers, private and governmental research institutions, and universities around the world are investing vast resources to improve the accuracy of such predictive tools.

Expansion of the simulation to include the transmission, drivetrain, tires and wheels, vehicle aerodynamics, rolling resistance, frictional losses, accessory loads, and the influence of control system response, calibration strategies, and hundreds of other parameters creates models of sufficient size to tax even high-power computers. Morever, such sophisticated models still require experimental verification and calibration and are best used to quantify incremental improvements on individual vehicle models. They also require the input of proprietary data.

The committee's charge was to estimate the potential for fuel economy improvements, not to define new regulatory standards. Hence it desired only a general understanding of the potential for fuel economy gain for different types of vehicles and what the relative costs might be. In addition, the committee wished to determine which technologies are currently being applied in markets where the high price of fuel provides an economic incentive for the introduction of new technology for reduced fuel consumption.

Although the committee is familiar with the state-of-the-art analytical methods identified above, it did not have the resources, time, or access to proprietary data necessary to employ such methods. Therefore it used a simpler methodology to provide approximate results. The committee identified candidate technologies, as explained in Chapter 3 of the prepublication version of the report, that could be considered for application in various types of vehicles. It then estimated ranges of possible improvements in fuel consumption and costs associated with these technologies. Finally, it assembled packages of technologies, deemed revelant to different vehicle classes, and estimated the total impact on fuel economy and costs. This approach allowed the committee to estimate potential changes in a wide variety of vehicle classes within the boundary conditions of the study. The committee notes that similar methods were used in the 1992 NRC analysis of automotive fuel economy potential (NRC, 1992) and by many studies in the published literature over the past 25 years (see, e.g., Greene and DeCicco, 2000, for a review).

Analytical Issues

Technical input to the study included a review of technical publications, a review of automotive manufacturer announcements of new technology introductions and reported fuel consumption (economy) benefits, and information acquired directly from automotive manufacturers and suppliers in the United States and abroad. The committee evaluated vehicle features (engine size, number of cylinders, state of technology) and published performance and fuel economy data. It assessed engine, transmission, and vehicle-related energy consumption, system losses, and potential improvements in thermal or mechanical efficiency. Finally, the committee applied engineering judgment to reduce an exceedingly complex and seemingly infinite number of possible technology combinations—and their relative performance, fuel consumption, drivability, production costs, and emissions compliance trade-offs—into a more manageable, though approximate, analysis.

Most of the technologies considered in the committee's analysis either are in, or will soon enter, production in the United States, Japan, or Europe. Promising emerging technologies, which are not completely developed but are sufficiently well understood, were also included. Background information concerning these technologies is given in Chapter 3 of the CAFE report. The potential choice of technologies differs by vehicle class and intended use. In addition, the ease of implementation into product plans and consumer-based preferences will influence whether a technology enters production at all.

The analysis was complicated by the need to infer potential fuel consumption benefits from published data in which experimental results were based on European (NEDC) or Japanese (10/11 mode) test cycles. Furthermore, differences in exhaust emission regulations, especially between European and U.S. Tier II or California standards, can have a great effect on the potential application of several technologies.[7]

The potential of each technology to improve fuel economy, and the costs of implementing the technology, were determined from the sources listed above. Both fuel economy (FE) benefits and costs are expressed in terms of a range, with low and high values, because of the uncertainty involved.[8] The benefit is expressed as a percent reduction of fuel consumption (FC; gallons/100 miles).

The fuel consumption ranges were adjusted in an attempt to account for potential double-counting of benefits. Attachment E shows how FC improvements were modified to avoid double-counting. It also shows that most of the technologies considered have primary and secondary benefits related to the reduction of different types of losses or improvements in thermal efficiency. In general, this strategy results in predicted improvements for individual technologies that are lower than the values commonly found in the literature.

In addition, subsequent to the release of the prepublication version, the committee simulated one case, the midsize SUV, in order to evaluate potential inaccuracies in its simplified methodology. This sample simulation is presented in Attachment C.

To assist in evaluating near-term potential (within 10 years) versus long-range predictions (10-15 years or beyond), the committee considered three technology paths with three different levels of optimism regarding technology implementation. The technologies grouped within these paths were chosen based on current production availability (in the United States, Europe, or Japan), general compatibility with the dominant vehicle attributes (engine size/power, transmission

---

[7] This is especially true in the case of lean combustion concepts (direct-injection diesel and gasoline), which are unlikely to penetrate U.S. markets rapidly due to production cost and emissions compliance issues, even though they are quickly approaching 50 percent of the new vehicle sales in Europe. The committee examined these technologies but did not include them in any of the paths because of high uncertainty concerning exhaust emissions compliance and production cost. Nevertheless, it is quite possible that one or more will be successful. In such a case, fuel economy levels higher than any of those estimated by the committee could become feasible.

[8] Note that the economic analysis in Chapter 4, including that in the prepublication version, heavily weights the average but statistically considers the uncertainty represented by the high and low values.

type, vehicle size, intended use, and so on), and, very qualitatively, "ease of implementation," related to possible product introduction. The committee believes that all these paths are plausible under some conditions, but it notes that technologies must meet development, production, customer acceptance, or other corporate boundary conditions that were not necessarily analyzed by the committee.

The baseline was determined from the EPA listings (EPA, 2001) for each class of vehicle, but FC was increased by 3.5 percent to account for potential weight increases (estimated at 5 percent) for future safety-specific and design changes. Aggregate reductions in FC were calculated for each path by simple multiplication of adjusted values as technologies germane to the path were added. Lower and upper bounds were calculated separately and an average was determined at each step. Aggregate costs were determined by adding the low and high costs separately and taking the average for each point, starting from zero. However, these costs were intended to approximate only the incremental cost to the consumer that could be attributed to improved fuel consumption alone. Box 1 shows some detailed calculations as an example.

Box 1
Example: Midsize SUV Path 1

The baseline fuel economy value is 21 mpg,[1] which is converted to fuel consumption (FC): 4.76 gallons/100 miles. The 3.5 percent penalty for safety equipment brings the starting point to 4.93 gallons/100 miles.

Six engine technologies, two transmission technologies, and one vehicle technology are included in Path 1, as shown in Table 3-2 (Attachment A). The average value for the first one, "Engine Friction Reduction," is 3 percent. Thus the average FC value with that technology added is 4.93 x 0.97 = 4.78. The next technology, "Low Friction Lubricants," is estimated at 1 percent, resulting in a value of 4.78 x 0.99 = 4.73. With all six engine technologies, the averaged consumption estimate is 4.19 gallons/100 miles. Including the transmission and vehicle technologies brings the total to 3.96 gallons/100 miles (or 25.3 mpg), as listed in Table 3-4 (Attachment A). The cost values in Table 3-2 are averaged and then simply added.

---

[1]EPA. 2001. *Light-Duty Automotive Technology and Fuel Economy Trends 1975 Through 2001*, EPA 420-R-01-008, September.

Overall Assessment

The committee's methodology is admittedly simplistic. Nevertheless, the committee believes it to be sufficiently accurate for the purposes of the study. The overall results are consistent with those of other analyses, such as those done by Energy and Environmental Analysis, Inc. (EEA's report to the committee is in the NRC public access file). In addition, two examples of the use of new technologies illustrate the potential to be gained (Attachment D).

Several changes and corrections of minor errors in the prepublication version of the report also have been made. The baseline fuel economy level for each vehicle is now taken directly from EPA published data (EPA, 2001) instead of a prepublication data set. Existing use of technology assumes that an entire class incorporates a technology when 50 percent or more of the class sales weighted average does. Two entries in Tables 3-1, 3-2, and 3-3 are corrected: (1) FC benefit value for variable valve timing (VVT) is now 2-3 percent (compared with 1-2 percent in the prepublication version), and (2) variable valve lift and timing (VVLT) is now 1-2 percent (formerly 3-8 percent). The values in the prepublication version had been adjusted on the assumption that the technologies would be coupled with transmission and downsized/supercharged engine configurations and should have been corrected when the methodology was changed. Also, some combinations of technologies have been corrected, having been carried forward in the spreadsheets (see Tables 3-1 to 3-3). Table 3-4 and Figures 3-4 through 3-13, incorporating these changes, are shown in Attachment A.

ANALYSIS OF COST-EFFICIENT FUEL ECONOMY LEVELS

The cost-efficient analysis (called the "break-even analysis" in Chapter 4 of the prepublication version but renamed the "cost-efficient analysis" in order to eliminate a source of confusion) depends on the results of the technical analysis. This section revises the cost-efficient results accordingly (these results are also in the final CAFE report). In addition, it provides an improved interpretation of the results.

Changes to the Cost-Efficient Analysis

The committee made several minor changes to the fuel economy improvement and cost data after the economic analysis was completed. The current analyses in Chapters 3 and 4 are now entirely consistent. In addition, the changes discussed above have been incorporated. An error affecting only midsize passenger cars was discovered in the computer program used to perform the cost-efficient calculations.[9] Finally, the committee corrected a minor methodological inconsistency in sequencing the application of technologies.[10] These changes collectively had only a small impact on the estimated cost-efficient fuel economy levels. The largest impact was due to the changes to the base fuel economy levels, as explained in the previous section.

The new cost-efficient fuel economy numbers are presented in Attachment B. In some cases, changes from the prepublication version tables exceed 2 mpg. These differences are due primarily to changes in the base miles per gallon (mpg) estimates. For example, the base mpg for large cars has been increased from 21.2 to 24.9 mpg. Chiefly as a result of this change, cost-

---

[9] The committee is grateful to Walt Kreucher of Ford Motor Company for pointing out this error.

[10] The method used to construct cost curves for fuel economy improvement begins by ranking technologies according to a cost-effectiveness index. The index is the midpoint of the range of "average" percent improvement divided by the midpoint of the range of "average" cost. Technologies are ranked from highest to lowest, in effect assuming that technologies will be implemented in order of cost-effectiveness. While this method is in accord with economic theory, it does not necessarily respect engineering reality. In one case, a technology (42-volt electrical system) that would have to be implemented before a second technology (integrated starter/generator) was ranked lower. This problem was solved by adding their costs and percent improvements as if they were one technology, in effect assuming that the two would be implemented simultaneously. The committee discovered this inconsistency prior to release of the report, determined that it did not significantly affect the calculations and, in the interests of a timely release of the prepublication version of the CAFE report, did not revise the method. The reconsideration of the report afforded the opportunity to make this revision.

efficient mpg levels are higher, but the percent increase is lower. Changes in the estimation method (described above) also contributed to changes in the estimates for cost-efficient fuel economy levels, but to a lesser degree.

Interpreting the Cost-Efficient Fuel Economy Estimates

The committee's analysis of cost-efficient (formerly referred to as "break-even") fuel economy levels has been more often misunderstood than understood. The committee acknowledges its responsibility to adequately explain this analysis and is taking this opportunity to clarify the meaning and proper interpretation of its analysis.

Cost-efficient fuel economy levels represent the point at which the cost of another small increment in fuel economy equals the value of the fuel saved by that increment. They do *not* represent the point at which the total cost of improving fuel economy equals the total value of the fuel saved. All mpg increases before this last increment more than pay for their cost in lifetime fuel savings. In general, the total value of fuel saved exceeds the total cost at the cost-efficient mpg level, quite significantly in some cases.

The originally published versions of Tables 4-2 and 4-3 included the costs but not the value of fuel savings, contributing to confusion about how to interpret the cost-efficient mpg levels. The revised tables, in Attachment B, show both total costs and the total value of fuel savings.

The committee emphasizes that these calculations depend on several key assumptions that are surrounded by substantial uncertainty. The analysis of cost efficiency considers only the consumer's costs and benefits. Societal benefits, such as external costs, are *not* reflected in the cost-efficient results (these costs are discussed in Chapter 5 of the report). From a societal perspective, higher or lower fuel economy levels might be preferred, depending on assumptions made regarding externality costs and how they should be applied.

Other critical assumptions concern the annual use and life of vehicles, consumers' perception of the value of fuel savings over the life of the vehicle, and the relationship between government fuel economy estimates and what motorists actually obtain in their day-to-day driving. As in the prepublication version, two sets of estimates are presented. One is based on fuel savings over the full expected life of a vehicle. The other is based on just the first 3 years of a vehicle's life.

REFERENCES

Environmental Protection Agency (EPA). 2001. *Light-Duty Automotive Technology and Fuel Economy Trends 1975 Through 2001,* EPA420-R-01-008, September.

Greene, David L., and John DeCicco. 2000. "Engineering-Economic Analyses of Automotive Fuel Economy Potential in the United States," *Annual Review of Energy and Environment.*

National Research Council (NRC). 1992. *Automotive Fuel Economy: How Far Should We Go?* Washington, D.C.: National Academy Press.

Sovran, G., and M.S. Bohn. 1981. *Formulae for the Tractive-Energy Requirements of Vehicles Driving the EPA Schedules.* SAE paper No. 810184, Detroit.

COMMITTEE ON THE EFFECTIVENESS AND IMPACT OF CORPORATE AVERAGE FUEL ECONOMY (CAFE) STANDARDS

PAUL R. PORTNEY, Chair, Resources for the Future, Washington, D.C.
DAVID L. MORRISON, Vice Chair, U.S. Nuclear Regulatory Commission (retired), Cary, North Carolina
MICHAEL M. FINKELSTEIN, Michael Finkelstein & Associates, Washington, D.C.
DAVID L. GREENE, Oak Ridge National Laboratory, Knoxville, Tennessee
JOHN H. JOHNSON, Michigan Technological University, Houghton, Michigan
MARYANN N. KELLER, priceline.com (retired), Greenwich, Connecticut
CHARLES A. LAVE, University of California (emeritus), Irvine
ADRIAN K. LUND, Insurance Institute for Highway Safety, Arlington, Virginia
PHILLIP S. MYERS, NAE, University of Wisconsin, Madison (emeritus)
GARY ROGERS, FEV Engine Technology, Inc., Auburn Hills, Michigan
PHILIP R. SHARP, Harvard University, Cambridge, Massachusetts
JAMES L. SWEENEY, Stanford University, Stanford, California
JOHN J. WISE, NAE, Mobil Research and Development Corporation (retired), Princeton, New Jersey

Project Staff

JAMES ZUCCHETTO, Director, Board on Energy and Environmental Systems (BEES)
ALAN CRANE, Responsible Staff Officer, Effectiveness and Impact of Corporate Average Fuel Economy (CAFE) Standards
PANOLA D. GOLSON, Senior Project Assistant, BEES

ACKNOWLEDGMENTS

The committee was aided by a consultant, K.G. Duleep of Energy and Environmental Analysis, Inc. He provided analyses to the committee, which the committee used in addition to the many other sources of information it received.

This letter report has been reviewed by individuals chosen for their diverse perspectives and technical expertise, in accordance with procedures approved by the Report Review Committee of the National Research Council (NRC). The purpose of this independent review is to provide candid and critical comments that will assist the authors and the NRC in making the published report as sound as possible and to ensure that the report meets institutional standards for objectivity, evidence, and responsiveness to the study charge. The content of the review comments and draft manuscript remain confidential to protect the integrity of the deliberative process. We wish to thank the following individuals for their participation in the review of this letter report:

Charles Amann (NAE), General Motors (retired),
Feng An, Consultant,
Francois Castaing (NAE), Castaing & Associates,
David E. Foster; University of Wisconsin,
Paul MacCready (NAE), AeroVironment, Inc.,
Craig Marks (NAE), Creative Management Solutions,
Steve Plotkin, Argonne National Laboratory,
Marc Ross, University of Michigan, and
Michael Walsh, Consultant.

Although the reviewers listed above have provided many constructive comments and suggestions, they were not asked to endorse the conclusions or recommendations, nor did they see the final draft of the report before its release. The review of this letter report was overseen by John Heywood (NAE), Massachusetts Institute of Technology, and Gerald P. Dinneen (NAE), Honeywell, Inc. (retired). Appointed by the National Research Council, they were responsible for making certain that an independent examination of the report was carried out in accordance with institutional procedures and that all review comments were carefully considered. Responsibility for the final content of this report rests entirely with the authoring committee and the institution.

Attachment A

Please see Chapter 3, Tables 3-1 to 3-4 and Figures 3-4 to 3-13.

Attachment B

Please see Chapter 4, Tables 4-2 and 4-3 and Figures 4-5 and 4-6.

Attachment C
Energy Balance Analysis

Written correspondence and verbal testimony offered during the October 5, 2001, public hearing suggested energy balance analysis as a method to avoid the potential for double counting of benefits. As noted above, such analyses can most accurately be applied to a particular vehicle with unique engine, powertrain, and vehicle characteristics and must be performed for each incremental improvement. In addition, engine map data (torque, speed, fuel consumption) must be incorporated that represent the incremental improvements for the various technologies being considered. These data must then be linked to transmission, driveline, and vehicular-specific parameters (rolling resistance, aerodynamic drag, etc.) and then simulated over various driving cycles. Furthermore, as the complexity of the technologies increases to include such things as intake valve throttling, integrated starter generators, or electrically controlled accessories, then assumptions and decisions on operating strategy, drivability, and emissions compliance are required that are time-consuming and require increasing amounts of proprietary data for simulation.

In an attempt to determine whether some fundamental flaws, resulting in gross errors, had inadvertently entered the judgment-simplified analysis described above, the committee conducted a simulation of a single vehicle (midsize SUV), for which it had access to data that could be used to attempt a more in-depth energy consumption/balance-type analysis. The analysis employed the computational methods outlined by Sovran and Bohn (SAE 810184) and a commercial engine simulation code (GT-Power[1]) to assess the contribution of gas exchange (pumping), thermodynamic efficiency (indicated efficiency), and cooling and exhaust heat losses. A proprietary test cycle simulation code was also employed to determine the percentage of fuel consumed by the simulated vehicle during deceleration and idle operation over the FTP-75 highway and city cycles, weighted (55/45), for a combined-cycle estimation.

Using the Sovran/Bohn equations, together with some limited cycle-estimated engine parameters (indicated efficiency, torque (brake mean effective pressure, BMEP), and speed (RPM)), the model was calibrated to 20.3 mpg for a simulated midsize SUV with a weight of 4300 lbs. A more detailed summary is shown in the table below. The committee's overall conclusion is that the path 1 and path 2 estimated average fuel consumption improvements in Table 3-4 (see Attachment A) appear quite reasonable, although the uncertainty in the analysis grows as more technology features are considered. The average path 3 prediction is, by definition, more aggressive. However, as previously stated, all Path 3 scenarios are presented as examples only, with increasing uncertainty as the number of technologies increases.

Overall, the committee believes that its judgment-based approach provides a sufficiently rigorous analysis for drawing the conclusions included in its report.

Reference

Sovran, G., and M.S. Bohn. 1981. Formulae for the Tractive-Energy Requirements of Vehicles Driving the EPA Schedules. SAE paper No. 810184, Detroit.

---

[1]Cycle analysis software marketed by Gamma Technologies, Inc., and used by nearly all major engine manufacturers worldwide in the design and development of new engines.

Category and Performance	Base Engine	Path 1	Path 2	Path 3	Aggressive Path 3	Upper Bound Case
Engine size (L)	4.0	3.5 (12.5% downsizing due to 2V to 4V)	3.4 (15% downsizing due to 2V to 4V,VVT, etc.)	3.0 (25% downsizing, due to 2V to 4V,VVT and supercharging)	2.5 (37.5% downsizing, due to 2V to 4V,VVT and supercharging)	2.3 (42.5% downsizing, due to 2V to 4V,VVT and supercharging)
FMEP (bar)	0.6 (city) and 0.65 (highway)	0.6 (city) and 0.65 (highway)	0.53 (city) and 0.57 (highway) (12% reduction)	0.54 (city) and 0.59 (highway) (10% reduction)	0.42 (city) and 0.46 (highway) (30% reduction)	0.35 (city) and 0.38(highway) (42% reduction)
PMEP (bar)	0.5 (city and highway)	0.4 for active cylinders (20% down) and 0.1 for deactivated cylinders (80% down)	0.4 for active cylinders (20% down) and 0.1 for deactivated cylinders (80% down)	0.10 (city and highway, 80% down, mainly due to eliminating intake throttle	0.05 (city and highway, 90% down, mainly due to eliminating intake throttle)	0.05 (city and highway, 90% down, mainly due to eliminating of intake throttle)
Indicated efficiency (%)	37	37	37.7 (2% increase from 2V to 4 V,VVLT and intake valve throttling)	38.9 (5% increase due to 4V and VCR)	38.9 (5% increase due to 4V and VCR)	39.2 (6% increase due to 4V and VCR)
Cylinder deactivation ?	No	Yes (3 cylinder deactivated)	Yes (3 cylinder deactivated)	No	No	No
Braking and idle-off ?	No	No	Yes	Yes	Yes	Yes
Vehicle loss reduction	No	-0.75% (weight increase-rolling resistance reduction, etc)	8% (account for averaged rolling /CD and accessory improvements, and 1/3 of transmission improvement, accessory improvement)	8.6% (account for averaged rolling /CD and accessory improvements, and 1/3 of transmission improvement, accessory improvement)	13% (account for max. rolling /CD and accessory improvements, and 1/3 of transmission improvement, accessory improvement)	13% (account for max. rolling /CD and accessory improvements, and 1/3 of transmission improvement, accessory improvement)
FE –city (mpg)	15.6	19.4 (24%)	25.6 (64%)	28.9 (85%)	33.8 (117%)	35.6 (128%)
FE –highway (mpg)	27.7	32.5 (18%)	37.3 (35%)	41.1 (48%)	47.1 (70%)	49.3 (78%)
FE – combined (mpg)	21.0	25.3 (20%)	30.9 (47%)	34.4 (64%)	39.8 (90%)	41.8 (99%)
FC – combined (g/100m)	4.76	3.95 (17%)	3.24 (32%)	2.92 (39%)	2.51 (47%)	2.40 (50%)
FE improvement (on base)	0	20%	47%	64%	90%	99%
FC improvement (on base)	0	16.9%	31.9%	38.8%	47.3%	49.6%
BSFC excluding idle + braking (g/kwh)	403	330	320	289	261	249
BSFC including idle + braking (g/kwh)	446	364	320	289	261	249

Attachment D
Examples of Technology Improvements

PAST IMPROVEMENTS

Over the past several years, DaimlerChrysler has introduced a new family of engines, replacing engines with push rods and overhead valves. One of them, the 2.7 liter engine, produces 200 horsepower, 40 more than its older equivalent, yet uses 10 percent less fuel. Had it been kept at the same power, fuel economy presumably could have been improved significantly more. Several technologies were used to achieve these improvements, including overhead cams (OHC), 4 valves/cylinder (the committee's estimated average contribution of these together is 3 percent), reduced engine size (6 percent when combined with a supercharger for even greater downsizing), and reduced friction losses (3 percent). According to the committee's values, these technologies would have contributed some 12 percent to fuel consumption reduction on average. Had DaimlerChrysler included variable valve timing and lift, fuel economy improvements of well over 20 percent might have been achieved for the same output. DaimlerChrysler used other factors, including electronic exhaust gas recirculation (EGR) which is tuned for high fuel economy as well as low emissions and lower idle speed, which were not included in the committee's list. This example is included to illustrate how much improvement is possible with redesign and technology upgrades.

FUTURE POTENTIAL

The variable compression ratio (VCR) engine, pioneered by Saab (which is now part of GM), is reported to achieve a 30 percent fuel consumption reduction (equivalent to a fuel economy gain of 43 percent) compared to a naturally aspirated conventional engine (GM, 2001). The Saab VCR uses a supercharger to achieve the same power as a larger engine. Most engines, in particular supercharged and turbocharged engines, must have a lower-than-optimal compression ratio to avoid knocking at high load. The compression ratio in the Saab engine can vary between 8 and 14. Variable compression allows the engine to work at high efficiency and produce a very high output (150 bhp/liter—more than twice the level of most current engines).

In the committee's analysis, all Path 3 engine technologies, production-intent and emerging including VCR, collectively result in an average fuel consumption reduction of 24 percent for midsize cars. Thus the committee's projection is conservative relative to Saab's 30 percent reduction, especially since the Saab VCR is presumably benchmarked against European engines that already have more of these sophisticated technologies than U.S. manufacturers use.

In the long term, even more innovative concepts may become available. For example, the hydraulic hybrid vehicle is drawing attention because it can store a much higher fraction of the energy from regenerative braking than can a HEV. One research group estimates that a car equivalent to a Ford Taurus could achieve 61 mpg in the urban driving cycle (Beachley and Fronczak, 1997).

Toyota is taking another approach, using a capacitor to store energy from regenerative braking. Toyota estimates that the four passenger prototype it is building will achieve 88 mpg (Toyota, 2001).

REFERENCES

Norman H. Beachley and Frank J. Fronczak, Advances in Accumulator Car Design, SAE 972645, Presented at the Future Transportation Technology Conference, San Diego, California, August 6-8, 1997.
General Motors Website. <www.gm.com/cgi-bin/pr_display.pl?1228>.
Toyota Website. <http://global.toyota.com/tokyomotorshow/es3/data.htm>.

Attachment E
Derivation of Fuel Consumption Improvement Values

	Engine Friction Reduction	FC Improvement	
		From Base	From Ref.[1]
Technology Description	Vehicle fuel consumption reduction resulting from reduced engine friction		
Primary Benefits	Higher brake mean effective pressure (BMEP) for the same indicated mean effective pressure (IMEP)		
Secondary Benefits	Higher BMEP allows engine downsizing.		
FC Improvement	Base: 2V baseline engine; Reference: 2V baseline engine	1 ~ 5 %	1 ~ 5 %
Example of Application	General technology to improve engine efficiency		
Reference	FEV, M. Schwaderlapp, F. Koch, J. Dohmen Fisita 2000-3, Seoul Congress Conclusions: In the next 10 years it will be possible to reduce SI engine fuel consumption by 8 – 13% through friction reduction		8 - 13 %

	Low Friction Lubricants	FC Improvement	
		From Base	From Ref.
Technology Description	Low friction lubricant to reduce engine friction and driveline parasitic losses.		
Primary Benefits	Low engine friction to reduce vehicle fuel consumption		
Secondary Benefits	Low friction to reduce driveline parasitic losses and vehicle fuel consumption		
FC Improvement	Base: 2V baseline engine; Reference: 2V baseline engine	1.0 %	1.0 %
Example of Application			
Reference	Toyota / Nippon Oil: K. Aklyama, T. Ashida; K. Inoue, E. Tominaga SAE-Paper 951037 Conclusion: Using additive in the lubricant oil reduces the fuel consumption by 2.7% for a 4.0L-V8-4V engine		2.7 %

---

[1] "Reference" refers to a vehicle with prior technologies already implemented. Thus it is the incremental improvement in a series of steps. It is lower than the base improvement (except for the first step in each category) to account for double-counting and other diminishing returns.

	Multivalve, Overhead Camshaft (2V vs. 4V)	FC Improvement	
		From Base	From Ref.
Technology Description	Improvement from 2 valve engine into a multiintake valve engine (including total of 3, 4, and 5 valves per cylinder)		
Primary Benefits	Lower pumping losses: larger gas exchange flow area Less friction: higher mechanic efficiency due to higher engine IMEP		
Secondary Benefits	Less pumping losses: engine down size with higher power density Higher thermal efficiency: higher compression ratio due to less knocking tendency and faster combustion process with central spark plug position		
FC Improvement	Base: 2V baseline engine; Reference: 2V baseline engine	2 ~ 5%	2 ~ 5%
Example of Application	Advanced engines from Ford, GM, and DC		
Reference	Volkswagen: R. Szengel, H. Endres 6. Aachener Kolloquium (1997) Conclusion: A 1.4L-I4-4V engine improves the fuel consumption by 11% (MVEG) in comparison to a 1.6L-I4-2V engine		11% FC in MVEG
	Ford: D. Graham, S. Gerlach, J. Meurer. SAE-Paper 962234 Conclusion: new valve train design (from OHV to SOHC) with 2 valves per cylinder plus additional changes (higher CR, less valve train moving mass) result in a 28% increase in power, 11% increase in torque and 4.5% reduction in fuel consumption (11.2 to 10.7 L/100km, M-H) for a 4.0L-V6-2V engine.		4.5% FC (OHV, 2V to SOHC, 2V) +28% power +11% torque
	Sloan Automotive Laboratory / MIT: Dale Chon, John Heywood SAE-Paper 2000-01-0565 Conclusion: The changing preference from 2-valve to 4-valve per-cylinder is a major factor of current engine power and efficiency improvement; the emergence of variable valve timing engines suggests a possible new trend will emerge.		

APPENDIX F

	Variable Valve Timing (VVT)	FC Improvement	
		From Base	From Ref.
Technology Description	Variable valve timing in the limited range through cam phase control		
Primary Benefits	Less pumping losses: later IVC to reduce intake throttle restriction for the same load		
Secondary Benefits	Less pumping losses: down size due to better torque compatibility at high and low engine speed for the same vehicle performance		
FC Improvement	Base: 2V baseline engine; Reference: 4V OHC engine	4 ~ 8 %	2 ~ 3 %
Example of Application	Toyota VVT-i; BMW Vanos		
Reference	Ford: R.A. Stein, K.M. Galietti, T.G. Leone SAE-Paper 950975 Conclusion: for a 4.6L-V8-2V engine in a 4,000 lb vehicle benefit in M-H fuel consumption of 3.2% with unconstrained cam retard and 2.8% (M-H) with constrained cam retard (10% EGR)		2.8 ~ 3.2% V8, 2V engine
	Ford: T.G. Leone, E.J. Christenson, R.A. Stein SAE-Paper 960584 Conclusion: for a 2.0L-I4-4V engine in a 3,125 lb vehicle benefit in M-H fuel consumption of 0.5-2.0% (10-15% EGR)		0.5-2.0% I4, 4V engine
	Toyota: Y. Moriya, A. Watanabe, H Uda, H. Kawamura, M. Yoshioka, M. Adachi. SAE-Paper 960579 Conclusion: for a 3.0L-I6-4V engine the VVT-i technology (phasing of intake valves) improved the fuel consumption by 6% on the 10-15 official Japanese mode.		6% Japanese mode I6, 4V engine
	Ford: D.L. Boggs, H.S. Hilbert, M.M. Schechter. SAE-Paper 950089 Conclusion: for a 1.6L-I4 engine the later intake valve closing improved the BSFC by 15% (10% EGR).		15% (BSFC) I4, Late IVC
	MAZDA / Kanesaka TI: T. Goto, K. Hatamura, S. Takizawa, N. Hayama, H. Abe, H. Kanesaka. SAE-Paper 940198 Conclusion: A 2.3L-V6-4V boosted engine with a Miller cycle (late intake valve closing) has a 10-15% higher fuel efficiency compared to natural aspiration (NA) engine with same maximum torque. 25% reduction in friction loss because of lower displacement. Expected 13% increase in fuel consumption of 2.3L Miller engine compared to 3.3L NA engine.		10 ~ 15% Fuel Efficiency, Miller cycle
	Mitsubishi: K. Hatano, K. Iida, H. Higashi, S. Murata. SAE-Paper 930878 Conclusion: A 1.6L-I4-4V engine reached an increase in fuel efficiency up to 16% (Japanese Test Driving Cycle) and an power increase of 20%.		Up to 16% in FC 20% Power
	Honda/Nissan/...: S. Shiga; S. Yagi; M. Morita; T. Matsumoto; H. Nakamura; T. Karasawa SAE-Paper 960585 Conclusion: For a 0.25L-I1-4V test engine an early closing of the intake valve results in up to 7% improvement in thermal efficiency		Up to 7% Fuel Efficiency
	Ricardo: C. Gray SAE-Pager 880386 Conclusion: Variable intake valve closing and cam timing duration improves part load fuel consumption by 3 ~ 5 %		3 ~ 5% at part load

Variable Valve Timing and Variable Valve Lift (VVLT)		FC Improvement	
		From Base	From Ref.
Technology Description	Valve lift and valve timing controlled according to engine load and speed, with step controlled mechanism		
Primary Benefits	Less pumping losses: partially use intake valve timing and lift control for intake throttle control Higher thermal efficiency: for better mixture formation with intake valve throttling		
Secondary Benefits	Less pumping losses: engine down size with higher power density		
FC Improvement	Base: 2V baseline engine; Reference: VVT engine	5 ~ 10%	1 ~ 2 %
Example of Application	Honda i-VTEC; Porsche Variocam Plus; Toyota VVLT-i		
Reference	Honda: M. Matsuki, K. Nakano, T. Amemiya, Y. Tanabe, D. Shimizu, I. Ohmura SAE-Paper 960583 Conclusion: for a 1.5L-I4-4V engine the 3-stages VTEC technology (three different cams) improved the power output by 40% with the same fuel consumption	40% more power with same fuel consumption	
	Porsche: C. Brüstle, D. Schwarzenthal. SAE-PAPER 980766 Conclusion: for a B6-4V engine the fuel consumption could be reduced by 3-9% with variable valve lift		3-9%
	Meta: P. Kreuter, P. Heuser, J. Reinicke-Murmann, R. Erz, U. Peter. SAE-Paper 1999-01-0329 Conclusion: For a 2.0L-I4-4V engine the VVLT system improved the fuel efficiency by 11% to 15% in idle speed	11% to 15% at idle	

Cylinder Deactivation		FC Improvement	
		From Base	From Ref.
Technology Description	Deactivate number of cylinders so that the active cylinders work on higher BMEP level, normally valve deactivation is necessary		
Primary Benefits	The active cylinders have less pumping loss with higher BMEP level		
Secondary Benefits			
FC Improvement	Base: 2V baseline engine; Reference: VVTL engine	8 ~ 16%	3 ~ 6%
Example of Application	Mercedes 5.0 L V8 and 6.0 L V12		
Reference	Meta: P. Kreuter, P. Heuser, J. Reinicke-Murmann, R. Erz, P. Stein, U. Peter. SAE-Paper 2001-01-0240 Conclusion: A I4 engine with cylinder valve deactivation (CVD) showed 20% improvement in fuel consumption at low engine speed. A V8 engine showed 6-8% improvement in fuel consumption for the New European Driving Cycle		6-8% FC in NEDC
	Daimler-Chrysler: M. Fortnagel, G. Doll, K. Kollmann, H.-K. Weining. MTZ 98 Sonderheft Conclusion: A 5.0L-V8-V3 engine has an improvement of 6.5% fuel consumption (New European Driving Cycle) and 10.3% in the FTP+HW cycle with the cylinder deactivation		6.5% FC in NEDC 10.3% FC in FTP+HW

Engine Accessory Improvement		FC Improvement	
		From Base	From Ref.
Technology Description	Improving the efficiency of accessory components or their power transmission to reduce the engine energy losses		
Primary Benefits	Direct reduction of vehicle fuel consumption		
Secondary Benefits	Higher net output allows engine downsizing		
FC Improvement	Base: 2V baseline engine; Reference: 4V OHC engine	3 ~ 7 %	1 ~ 2 %
Example of Application	Less coolant flow rate, less oil flow rate		
Reference	"Technology and Cost of Future Fuel Economy Improvements for Light-Duty Vehicles – Draft Final Report"; Energy and Environmental Analysis, Inc. – NAS Report – June 4, 2001 Conclusions: Between 0.5 and 1% reduction in fuel economy is possible		0.5 – 1 % reduction in fuel economy

Supercharging and Downsizing		FC Improvement	
		From Base	From Ref.
Technology Description	Reduce the engine displacement and supercharge it for the required power		
Primary Benefits	Less pumping loss at low load conditions; less friction power loss at the same FMEP; less Idle losses		
Secondary Benefits	None		
FC Improvement	Base: 2V baseline engine; Reference: 4V OHC engine	7 ~ 12 %	5 ~ 7 %
Example of Application			
Reference	FEV, Peter Walzer, 00ELE028 Future Engines For Cars Conclusions: Engine down size from 3L to 1.5L with supercharging and VCR, part load specific fuel consumption improves by 25%	25% at part load, with VCR	

5-Speed Automatic Transmission		FC Improvement	
		From Base	From Ref.
Technology Description	Added ratio places engine in better average speed/load operating point. Improvements in torque converter lockup via Slip Controlled Converter Clutch. Improved internal oil pump losses by reducing pressure. Closed-loop shift strategy. Reduction of gear drag losses. General weight reduction.		
Primary Benefits	Less pumping loss at low load conditions; less friction power loss at the same FMEP; lower Idle losses		
Secondary Benefits	Improved transmission efficiencies		
FC Improvement	Baseline: 4-speed; Reference: 4-speed	2 ~ 3 %	2 ~ 3 %
Example of Application			
Reference	SAE- 970689, "ZF 5-Speed Transmissions for Passenger Cars"; Heribert Scherer, Georg Gierer Auto 2000, "ZF 5-Speed Automatic Transmission"; Heribert Scherer Conclusions: A 5% reduction can be attributed to the new 5-speed transmission		5% on combined M-H FTP-75

Continuously Variable Transmission (CVT)		FC Improvement	
		From Base	From Ref.
Technology Description	Added ratio places engine in better average speed/load operating point. Elimination of torque converter with an optimized starting clutch procedure. Reduced work loss in the drive train and accessories due to the gear ratio characteristics unique to the CVT		
Primary Benefits	Less pumping loss at low load conditions; less friction power loss at the same FMEP; lower Idle losses		
Secondary Benefits	Improved drive train and accessory losses		
FC Improvement	Baseline: 4-speed, Reference: 5-speed	6 ~ 11 %	4 ~ 8 %
Example of Application	Audi A4 – Multitronic		
Reference	ATZ 8&9/2000, "Multitronic – The New Automatic Transmission from Audi – Parts 1 & 2"		
	SAE 970685, "ECOTRONIC – Continuously Variable ZF Transmission (CVT);" Manfred Boos and Herbert Mozer		
	SAE 1999-01-0754, "Development of an Engine-CVT Integrated Control System;" S. Sakaguchi, E. Kimura, K. Yamamoto Conclusions: A 9.3% reduction can be attributed to the CVT transmission		9.3% on MVEG

Aggressive Shift Logic		FC Improvement	
		From Base	From Ref.
Technology Descriptions	Improvements in torque converter lockup. Closed-loop shift control strategy		
Primary Benefits	Reduced transmission losses		
Secondary Benefits	None		
FC Improvement	Baseline: 4-speed, Reference: 5-speed	3 ~ 6 %	1 ~ 3 %
Example of Application			
Reference	"Technology and Cost of Future Fuel Economy Improvements for Light-Duty Vehicles – Draft Final Report"; Energy and Environmental Analysis, Inc. – NAS Report – June 4, 2001 Conclusions: A 9%-9.3% reduction can be attributed to aggressive shift logic with a 5-speed transmission		9.0-9.3 % improvement in Fuel Economy

6-Speed Automatic Transmission		FC Improvement	
		From Base	From Ref.
Technology Description	Added ratio places engine in better average speed/load operating point. Improved gearbox efficiency with outstanding direct drive efficiency and reduced gear drag losses. Improved internal oil pump losses by internally geared wheel-pump and improved volumetric efficiency and reduced leakage losses. Optimized oil supply with reduced leakage in the hydraulic controls and gearbox.		
Primary Benefits	Less pumping loss at low load conditions; less friction power loss at the same FMEP; less idle losses		
Secondary Benefits	Improved transmission efficiencies		
FC Improvement	Baseline: 4-speed, Reference: 5-speed	3 ~ 5 %	1 ~ 2 %
Example of Application	BMW 7-Series		
Reference	ATZ 9/ 2000, "6-Speed Automatic Transmission for the New BMW 7-Series;" Wolfgang Hall, Christian Bock Conclusions: A 5% reduction can be attributed to the new 6-speed transmission		5% on combined M-H FTP-75

Aerodynamic Drag Reduction		FC Improvement	
		From Base	From Ref.
Technology Description	Aerodynamic drag reduction via vehicle shape changes or reduced frontal area		
Primary Benefits	Reduced higher speed engine load required		
Secondary Benefits	None		
FC Improvement	Baseline: conventional vehicles; Reference: conventional vehicles	1 ~ 2 %	1 ~ 2 %
Example of Application			
Reference	"Technology and Cost of Future Fuel Economy Improvements for Light-Duty Vehicles – Draft Final Report"; Energy and Environmental Analysis, Inc. – NAS Report – June 4, 2001 Conclusions: A 10% drag reduction is possible with a result in 1.6 – 2.2 % FE reduction.		1.6 to 2.2% fuel economy reduction

Improve Rolling Resistance		FC Improvement	
		From Base	From Ref.
Technology Description	Reduced bearing, brake and driveline rotating forces. Improvements in tire rolling resistances through new tread designs and tire carcass improvements		
Primary Benefits	Reduced engine load required over entire speed range		
Secondary Benefits	None		
FC Improvement	Baseline: conventional vehicles; Reference: conventional vehicles	1 ~ 1.5 %	1 ~ 1.5 %
Example of Application			
Reference	"Technology and Cost of Future Fuel Economy Improvements for Light-Duty Vehicles – Draft Final Report"; Energy and Environmental Analysis, Inc. – NAS Report – June 4, 2001 Conclusions: A 10% rolling resistance reduction is possible with a result in 1.5 – 2.0 % FE reduction		1.6 to 2.2% fuel economy reduction

Safety Weight Increase		FC Improvement	
		From Base	From Ref.
Technology Description	Added weight to account for anticipated future safety structure, equipment or other features		
Primary Benefits	Increased engine load required		
Secondary Benefits	None		
FC Improvement	Baseline: conventional vehicles, Reference: conventional vehicles	-3 ~ -4 %	-3 ~ -4 %
Example of Application			
Reference	"Technology and Cost of Future Fuel Economy Improvements for Light-Duty Vehicles – Draft Final Report"; Energy and Environmental Analysis, Inc. – NAS Report – June 4, 2001 Conclusions: 10% weight reduction results in 6.6 to 8% reduction in FE. With a safety weight increase of 5% the committee used 3 to 4% FE reduction to account for this.		3 to 4% increase

Intake Valve Throttling		FC Improvement	
		From Base	From Ref.
Technology Description	Electronic or hydraulically controlled, mechanically actuated continuous variable valve timing and lift		
Primary Benefits	Less pumping losses: much less, or no, intake throttling for load control. Higher thermal efficiency: better mixture formation with intake valve throttling. Less friction: higher mechanical efficiency due to higher engine IMEP.		
Secondary Benefits	Less pumping losses: engine down size with higher power density		
FC Improvement	Base: 2V baseline engine; Reference: VVT engine	8 ~ 16%	3 ~ 6 %
Example of Application	BMW Valvetronic		
Reference	MTZ 10 2001, pp. 826-835 Conclusion: Valvetronic creates a fuel consumption reduction of 12% part load; 20% in idle; 14% reduction of fuel consumption for MVEG III compared to its predecessor.	20% idle 12% part load 14% MVEG III	
	Delphi: R.J Pierik, J.F. Burkhard SAE Paper 2000-01-1221 Conclusion: demonstrated brake specific fuel consumption (BSFC) of 12% at idle, 7-10% at low middle load, and 0-3% at middle to high load.		Idle: 12% low: 7% mid: 10% high: 0-3% (BSFC)
	Hyundai / Siemens: J. Lee, Ch. Lee, J.A. Nitkiewicz SAE-Paper 950816 Conclusion: For a 2.0L DOHC engine the fuel efficiency could be increased by 30% in idle; 3-4% in low speed; 5% in part load with "lost motion" technology. It uses conventional cam and create lost motion with hydraulic mechanism.		Idle: 30% low: 3-4% part load: 5% high: 0% torque: 9.8%
	BMW: R. Fierl, M. Klüting SAE-Paper 2000-01-1227 Conclusion: The electromechanical valve train offers a reduction in fuel consumption by about 10% plus 5% higher peak torque.		10%
	Nissan: S. Takemura, S. Aoyama, T. Sugiyama, T. Nohara, K. Moteki, M. Nakamura, S. Hara SAE-Paper 2001-01-0243 Conclusion: A variable actuation system showed fuel consumption of nearly 10%		10%
	University of Bucharest: N. Negurescu, C. Pana, M.G. Popa, A. Racovitza SAE-Paper 2001-01-0671 Conclusion: For a one-cylinder test engine VVT increases the efficiency by 10 to 29%		10-29% efficiency

Camless Valve Actuation		FC Improvement	
		From Base	From Ref.
Technology Description	Completely variable valve timing controlled and actuated by electromagnetic or high-pressure hydraulic means		
Primary Benefits	Less pumping losses: completely eliminate intake throttling valve for load control Higher thermal efficiency: higher compression ratio with less knocking tendency; better mixture formation with intake valve throttling Less friction: less valve train friction; higher mechanical efficiency due to higher engine IMEP		
Secondary Benefits	Less pumping losses: engine down size with higher power density		
FC Improvement	Base: 2V baseline engine; Reference: VVT engine	10 ~ 20%	5 ~ 10%
Example of Application	FEV EMV; Siemens EVT		
Reference	FEV: M. Pischinger, W. Salber, F. van der Staay, H. Baumgarten, H. Kemper FISITA – Seoul 2000 Conclusion: a reduction of 16% fuel consumption can be achieved by using the EMV-technology in a 1.6L-I4-4V engine	16% with EMV	

Variable Compression Ratio (VCR)		FC Improvement	
		From Base	From Ref.
Technology Description	Using higher compression ratio at low load condition for high thermal efficiency and low compression ratio at high load conditions to avoid knocking. Normally applies to supercharged-down size engines.		
Primary Benefits	Higher thermal efficiency at part load conditions		
Secondary Benefits	None		
FC Improvement	Base: 2V baseline engine; Reference: 4V OHC engine and supercharge down sizing	9 ~ 18%	2 ~ 6 %
Example of Application	SAAB VCR engine		
Reference	Saab: H. Drangel, L. Bergsten Aachen Kolloquium 2000 Conclusion: With the combination VCR / high charging and downsizing of the engine, it was possible to get the same power out of an 1.6L-I5-4V engine as a 3.0L-V6 engine. The resulting fuel consumption reduction is 30%		30%
	Daimler-Benz: F.G. Wirbeleit, K. Binder, D. Gwinner SAE-Paper 900229 Conclusion: In a V8 a VCR between 8 to 13.9:1 depending on the engine speed, the fuel consumption improves by 4% to 8%		4% - 8%
	Ford/University of Dar es Salaam: T. H. Ma, H. Rajbu SAE-Paper 884053 Conclusion: At 1,500 rpm and 2 bar BMEP condition, VVT alone achieves 8% BSFC; VVT+VCR achieves 19%		11% BSFC (1,500 rpm and 2 bar BMEP)

Automated Shift Manual Transmission		FC Improvement	
		From Base	From Ref.
Technology Descriptions	Improved gearbox efficiency with improved efficiency and reduced gear drag losses. Elimination or significant reductions of internal oil pump losses.		
Primary Benefits	Improved transmission efficiency		
Secondary Benefits	None		
FC Improvement	Baseline: 4-speed, Reference: 6-speed	6 ~ 10 %	3 ~ 5 %
Example of Application			
Reference	SAE Toptec – Modern Advances in Automatic Transmission Technology, "EMAT – Electro-Mechanical Automatic Transmission"; D. Carriere, J. Cherry, R. Reed, Jr. Conclusions: An estimated 10% improvement in fuel efficiency with improved performance		Estimated 10% Improvement in fuel efficiency

Advanced CVT's (Allows Higher Torque)		FC Improvement	
		From Base	From Ref.
Technology Description	Improved transmission efficiency using toroidal-shape and roller elements and special traction fluids. Permits use in higher torque applications.		
Primary Benefits	Improved transmission efficiency. Brings CVT to higher torque applications.		
Secondary Benefits	None		
FC Improvement	Baseline 4-speed; Reference: CVT	6 ~ 13 %	0 ~ 2 %
Example of Application			
Reference	Mazda's Future – Cars and Technology for Tomorrow Conclusions: A 20% improvement in fuel economy in the Japanese 10-15 mode compared with a current 4-speed automatic transmission	20% Improvement in fuel economy	

Integrated Starter Generator with Idle Off		FC Improvement	
		From Base	From Ref.
Technology Description	Integrated starter generator (ISG) cuts off fuel supply at idle and when the brakes are applied. Greater starter power enables the engine to be started immediately at higher speed.		
Primary Benefits	Less fuel loss when engine power is not necessary		
Secondary Benefits	None		
FC Improvement	Base: 2V baseline engine; Reference: 4V OHC engine	6 ~ 12 %	4 ~ 7 %
Example of Application			
Reference	"Technology and Cost of Future Fuel Economy Improvements for Light-Duty Vehicles – Draft Final Report"; Energy and Environmental Analysis, Inc. – NAS Report – June 4, 2001 Conclusions: Technology will provide for idle off, launch assist, improved power generation with a 9% – 11% FE improvement.		9 to 11% FE improvement

42 V Electrical System		FC Improvement	
		From Base	From Ref.
Technology Descriptions	Changing the vehicle operation voltage from 12V into 42V permitting electronically controlled thermal management (water pump). Enabling technology for 42V ISG.		
Primary Benefits	Less electrical power losses with less current flow through wires; higher efficiency of the electrical components		
Secondary Benefits	Enables higher efficiency ISG systems		
FC Improvement	Base: 2V baseline engine; Reference: 4V OHC engine	3 ~ 7 %	1 ~ 2 %
Example of Application			
Reference	"Wards Engine and Vehicle Technology Update," June 15, 2001, p. 7 Conclusions: Potential for electronic thermal management is 5% FE		5% FE improvement

Electric Power Steering		FC Improvement	
		From Base	From Ref.
Technology Description	Using electric motor to drive power steering		
Primary Benefits	Reduced parasitic losses due to optimized operation (only when needed)		
Secondary Benefits	None		
FC Improvement	Base: 2V baseline engine; Reference: 4V OHC engine	3.5 ~ 7.5 %	1.5 ~ 2.5 %
Example of Application			
Reference	ZF Lenksysteme: D. Peter, R. Gerhard SAE-Paper 199-01-0401 Conclusion: Reduction of fuel consumption by 2-3% by using electrical power steering instead of hydraulic power steering for a medium-sized vehicle.		2 - 3%